AF581859

Selected papers of the "1st International Conference on Nanofluids (ICNf)"

Selected papers of the "1st International Conference on Nanofluids (ICNf)"

Editors

Patrice Estellé

Leonor Hernández López

Matthias H. Buschmann

MDPI • Basel • Beijing • Wuhan • Barcelona • Belgrade • Manchester • Tokyo • Cluj • Tianjin

Editors

Patrice Estellé
Université de Rennes 1
France

Leonor Hernández López
Universitat Jaume I
Spain

Matthias H. Buschmann
Institut für Luft- und
Kältetechnik gGmbH Dresden
Germany

Editorial Office
MDPI
St. Alban-Anlage 66
4052 Basel, Switzerland

This is a reprint of articles from the Special Issue published online in the open access journal *Energies* (ISSN 1996-1073) (available at: https://www.mdpi.com/journal/energies/special_issues/ICNf2019).

For citation purposes, cite each article independently as indicated on the article page online and as indicated below:

LastName, A.A.; LastName, B.B.; LastName, C.C. Article Title. *Journal Name* **Year**, *Article Number*, Page Range.

ISBN 978-3-03936-662-0 (Hbk)
ISBN 978-3-03936-663-7 (PDF)

Contents

About the Editors

Patrice Estellé Associate Professor at the Université Rennes 1, (Rennes, France). His research interests currently include the thermophysical properties of nanofluids, nanofluid heat and mass transfer, and the flow properties of soft matter and suspensions.

Leonor Hernández López is Associate Professor at the University Jaume I (Castelló, Spain). She chairs the European Nanouptake COST action "Overcoming Barriers to Nanofluid Market Uptake" and has also chaired the ICNf2019 congress. Her research interests include nanofluid synthesis and thermophysical characterization, the improvement of heat transfer and energy storage in energy applications, and the development and optical characterization of solar nanofluids for optimizing solar radiation absorption

Matthias H. Buschmann is Project Manager at Institut für Luft- and Kältetechnik Dresden gGmbH (Dresden, Germany). His research interests are in the fields of heat transfer, turbulence theory, and air conditioning systems.

Editorial

Special Issue of the 1st International Conference on Nanofluids (ICNf19)

Patrice Estellé [1,*], Leonor Hernández López [2] and Matthias H. Buschmann [3]

1 Laboratoire de Génie Civil et Génie Mécanique(LGCGM), Université de Rennes, F-35000 Rennes, France
2 Departamento de Ingeniería Mecánica y Construcción, Universitat Jaume I, 12071 Castelló de la Plana, Spain; lhernand@uji.es
3 Institut für Luft- und Kältetechnik gGmbH Dresden, 01309 Dresden, Germany; Matthias.Buschmann@ilkdresden.de
* Correspondence: patrice.estelle@univ-rennes1.fr; Tel.: +33-223234200

Received: 26 March 2020; Accepted: 1 May 2020; Published: 6 May 2020

Abstract: This editorial note is dedicated to the 1st International Conference on Nanofluids (ICNf19), which was organized under the auspices of Nanouptake COST Action in June 2019, in Castelló (Spain). After a brief report about the conference issues, the successful selected contributions to this Special Issue of *Energies* about the ICNf19 are introduced.

Keywords: nanouptake; nanofluids; properties; energy; applications

---

1. Introduction

Since the early 1990s, nanofluids have received growing attention due to their strong potential as heat transfer fluids in energy systems. This interest is part of a wider context of efficient and clean energy demand. Nanofluid science is multidisciplinary and far-reaching, as many complex phenomena are involved and thus require knowledge in many scientific fields, leading to cross-practice collaborations.

In this context, four years ago, the Nanouptake CA15119 project [1] was launched by Enrique Julià Bolivar, under the auspices of the European Cooperation in Science and Technology (COST). The main objectives of this scientific project were to create a Europe-wide network of leading R+D+i institutions and of key industries, to reinforce existing collaborations at a European level to develop and foster the use of nanofluids as advanced heat transfer/thermal storage materials to increase the efficiency of heat exchange and storage systems [1].

With this goal, our European consortium is developing a common understanding of nanofluid preparation and characterization, addressing nanofluids' key barriers from the market and creating an open space for new relevant research.

This editorial note gives a short report of the 1st International Conference on Nanofluids and the 2nd European Symposium of Nanofluids (ICNf2019) [2] which we organized; a series of international conferences initiated by our consortium in 2017 with the 1st European Symposium of Nanofluids (ESNf2017) in Lisbon [3–5]. Then, the successful selected contributions to this Special Issue of *Energies* are introduced.

2. ICNf2019: A Short Report

ICNf2019 [2] was held in Castelló (Spain) from 26th to 28th June 2019, after the successful previous experience of the 1st European Symposium of Nanofluids (ESNf2017) two years before in Lisbon [3–5]. The new edition of the conference in 2019 started with the idea of expanding its reach outside Europe, in order to encourage the participation of speakers on an international scale.

The conference became a great opportunity to exchange knowledge among academics, researchers and companies, and also to share the current progress in the field of nanofluids for energy applications.

During the three days of the conference, and due to the number of important participants, three parallel sessions were required to cover the nine different nanofluids thematic sessions, ranging from basic research to real world industrial applications.

This first international edition of the conference has been a great success, both at the quantitative level (the number of communications and participants) and from a qualitative point of view (the scientific level and internationalization). More than 150 contributions have been received and, after a peer-review process from two different members of the Scientific Committee, 126 were accepted; 84 as oral and 42 as poster presentations. More than 200 participants from 45 different countries from five continents attended the conference. Five international keynote speakers from China, Australia, South Africa, Thailand and Germany, respectively, presented their high impact research in the plenary sessions of the congress.

All accepted contributions have been published in open access in the conference proceedings, which are freely available here [6]. Additionally, selected authors were invited to send extended versions of their conference contributions to the *Energies* ICNf2019 Special Issue. After a standard peer-review process, nine papers have been accepted for publication in this Special Issue. Before a short introduction of these papers, we wish to once more thank the people involved in the organization and the scientific committee which made this conference a great event, as well as all the sponsors, exhibitors, collaborators and partners.

3. ICNf2019: About the Selected Papers of the SI

Among the papers submitted to this special issue, nine articles have been published after careful review: seven full research papers and two reviews. Authors' geographical distribution (published papers) is:

- Spain (4);
- Bosnia and Herzegovina (2);
- Portugal (1);
- Slovakia (1);
- France (1)*;
- Turkey (1)*.

* Same submission with two correspondences.

The papers in this special issue cover a broad range of fundamental and applied research aspects on nanofluid science and development, and reflect the current investigations and knowledge in different topics. It should be mentioned that most of these papers also reflect the collaborations between researchers and labs involved in COST Action Nanouptake; some of these research works have been performed during Short-Term Scientific Missions (STSM), which have been one of the most successful aspects of this program.

The first paper, by Cabaleiro et al. [7], reports a comprehensive experimental characterization of phase change material (paraffin-in-water) nano-emulsions. The effect of dispersed phase concentrations of 2, 4 and 10 wt.% on the dynamic viscosity, surface tension and wettability of nano-emulsions is investigated, showing the impact of paraffin droplets and surfactant on these properties.

In [8], Grosu et al. propose two strategies to solve the creeping problem in thermal measuring systems for molten salts and molten salts based nanofluids. This paper shows that the use of crucibles with nanoscale roughness on the inner surface, or made/coated with a low surface energy material, allows the creeping phenomenon to be solved, controlling the wettability of the surfaces, and leads to proper measurement and property evaluation.

Regarding the wettability property of nanofluids, Çobanoğlu et al. [9] study the validity of single-phase contact angle correlations for several nanofluids with different types of nanoparticles dispersed in water. It is shown that these models can be used for dilute nanofluids at ambient conditions, and the authors also demonstrate that the droplet shapes of nanofluids are well predicted by a model based on force balance.

Rajnak et al. [10] evaluate the thermophysical properties, such as thermal conductivity, dielectric permittivity and viscosity, of magnetic nanofluid containing iron oxide nanoparticles and commercial oil. The influence of nanoparticle content and magnetic field on these properties is studied, showing in particular an anisotropy effect and a potential application of these nanofluids in power transformers.

In the paper from Martínez-Merino et al. [11], nanofluids based on one-dimensional MoS_2, WS_2 nanosheets and thermal oil currently used in concentrating solar plants (CSP) are prepared and characterized. The influence of nanostructure morphologies on nanofluid stability and thermal properties are presented and discussed. It is finally shown that the thermal conductivity of these nanofluids is strongly improved in comparison to thermal oil, leading to potential efficiency in solar collectors.

Two other papers deal with numerical simulations of nanofluid flow and heat transfer properties. In the first one, the laminar flow of ethylene glycol-based silicon nitride (EG-Si_3N_4) nanofluid in a smooth horizontal pipe subjected to forced heat convection with constant wall heat flux is numerically investigated by Berberović and Bikić [12], considering the thermophysical properties measured experimentally and the non-Newtonian features of nanofluids. An increase in heat transfer of the EG-Si_3N_4 nanofluid is reported in comparison to the EG for all the nanoparticle loadings and flow rates, evidencing a potential of these nanofluids for heat transfer applications.

The second article, from Alic [13], analyzes the entransy flow dissipation rate of three serially-connected cylindrical heating elements with Al_20_3 nanofluids as working fluids varying the concentration and flow rate. A model is developed based on a newly introduced dimension irreversibility ratio, defined as the ratio between entransy flow dissipation and thermally generated entropy, to optimize the geometric and process parameters of cylindrical heating elements.

The last two papers give an overview of the thermal and electroactive properties of graphene nanofluids [14], and both practical and theoretical recommendations about the measurement of the thermal conductivity of ionic melts and nanofluids [15], respectively. In [14], the authors focus on the development of graphene and reduced graphene oxide dispersed in aqueous and organic electrolytes, and the evaluation of their thermal and electrochemical properties. These nanofluids are demonstrated to be good candidates for heat transfer and energy storage applications.

Finally, in [15], the available techniques for the measurement of the thermal conductivity of fluids, including nanofluids, ionic liquids and molten salts, are critically analyzed, with special focus on transient methods.

4. Conclusions

This Editorial note was dedicated to the 1st International Conference on Nanofluids and the 2nd European Symposium of Nanofluids (ICNf2019). After a short report about the conference, the successful invited submissions [7–15] to this Special Issue of *Energies* were introduced. We believe that the articles in this special issue, whose main parts were presented during the conference, will contribute significantly to the development of nanofluid science and challenges in heat transfer and energy applications.

We would like to thank all the authors for their contributions in this special issue, and thank the editorial staff for their help during the process, as well as all the reviewers for their efforts in the evaluation of the submitted papers and their critical and meaningful comments. We found the selection and editing of the papers for this special issue to be very inspiring and rewarding.

In addition, this editorial note gives us the opportunity and pleasure to acknowledge all the members of the Core Group and Grant Holder Managers for their strong involvement in the achievement of this joint program. This Special Issue is dedicated to Enrique Julià Bolivar, who initiated this Action; nothing would have been possible without him.

Author Contributions: P.E., L.H.L. and M.H.B.; writing–review and editing. L.H.L. is the Nanouptake COST Action chair and chair of the conference. M.H.B. is a Working Group leader of the Nanouptake COST Action and vice-chair of the conference. P.E. is a member of the scientific committee of the conference, vice-chair of the conference session and the STSM coordinator of the Nanouptake COST Action. All authors have read and agreed to the published version of the manuscript.

Funding: The APC of papers [8,10–15] was funded by the COST Action Nanouptake (CA15119).

Acknowledgments: The conference and this editorial note are based upon works and activities from COST Action Nanouptake (CA 15119), supported by COST (European Cooperation in Science and Technology). COST is a funding agency for research and innovation networks. The financial support of COST, which is supported by the EU Framework Programme Horizon 2020, is gratefully acknowledged.

Conflicts of Interest: The authors declare no conflict of interest.

References

1. Nanouptake. Available online: http://www.nanouptake.eu/ (accessed on 1 May 2020).
2. 1st International Conference on Nanofluids (ICNf) and the 2nd European Symposium on Nanofluids (ESNf). Available online: http://icnf2019.com/index.php (accessed on 1 May 2020).
3. 1st European Symposium on Nanofluids (ESNf). Available online: http://esnf2017.campus.ciencias.ulisboa.pt/ (accessed on 1 May 2020).
4. Murshed, S.M.S.; Nieto de Castro, C.A.; Juliá Bolívar, J.E. Report on First European Symposium on Nanofluids. *Appl. Rheol.* **2018**, *28*, 45–47. [CrossRef]
5. Murshed, S.M.S. A Special Note on the First European Symposium on Nanofluids, NanoUptake, and the Memory of Professor José Enrique Juliá Bolívar. *J. Nanofluids* **2018**, *7*, 1033–1034. [CrossRef]
6. Available online: http://repositori.uji.es/xmlui/handle/10234/183448?locale-attribute=en (accessed on 1 May 2020).
7. Cabaleiro, D.; Hamze, S.; Agresti, F.; Estellé, P.; Barison, S.; Fedele, L.; Bobbo, S. Dynamic Viscosity, Surface Tension and Wetting Behavior Studies of Paraffin–in–Water Nano–Emulsions. *Energies* **2019**, *12*, 3334. [CrossRef]
8. Grosu, Y.; González-Fernández, L.; Nithiyanantham, U.; Faik, A. Wettability Control for Correct Thermophysical Properties Determination of Molten Salts and Their Nanofluids. *Energies* **2019**, *12*, 3765. [CrossRef]
9. Çobanoğlu, N.; Karadeniz, Z.H.; Estellé, P.; Martínez-Cuenca, R.; Buschmann, M.H. Prediction of Contact Angle of Nanofluids by Single-Phase Approaches. *Energies* **2019**, *12*, 4558. [CrossRef]
10. Rajnak, M.; Wu, Z.; Dolnik, B.; Paulovicova, K.; Tothova, J.; Cimbala, R.; Kurimský, J.; Kopcansky, P.; Sunden, B.; Wadsö, L.; et al. Magnetic Field Effect on Thermal, Dielectric, and Viscous Properties of a Transformer Oil-Based Magnetic Nanofluid. *Energies* **2019**, *12*, 4532. [CrossRef]
11. Martínez-Merino, P.; Alcántara, R.; Aguilar, T.; Gallardo, J.J.; Carrillo-Berdugo, I.; Gómez-Villarejo, R.; Rodríguez-Fernández, M.; Navas, J. Stability and Thermal Properties Study of Metal Chalcogenide-Based Nanofluids for Concentrating Solar Power. *Energies* **2019**, *12*, 4632. [CrossRef]
12. Berberović, E.; Bikić, S. Computational Study of Flow and Heat Transfer Characteristics of EG-Si3N4 Nanofluid in Laminar Flow in a Pipe in Forced Convection Regime. *Energies* **2020**, *13*, 74. [CrossRef]
13. Alic, F. Entransy Dissipation Analysis and New Irreversibility Dimension Ratio of Nanofluid Flow Through Adaptive Heating Elements. *Energies* **2020**, *13*, 114. [CrossRef]
14. Rueda-García, D.; Rodríguez-Laguna, M.R.; Chávez-Angel, E.P.; Dubal, D.; Cabán-Huertas, Z.; Benages-Vilau, R.; Gómez-Romero, P. From Thermal to Electroactive Graphene Nanofluids. *Energies* **2019**, *12*, 4545. [CrossRef]
15. Nieto de Castro, C.A.; Lourenço, M.J.V. Towards the Correct Measurement of Thermal Conductivity of Ionic Melts and Nanofluids. *Energies* **2020**, *13*, 99. [CrossRef]

Article

Dynamic Viscosity, Surface Tension and Wetting Behavior Studies of Paraffin–in–Water Nano–Emulsions †

David Cabaleiro [1,2,3,*], Samah Hamze [1], Filippo Agresti [4], Patrice Estellé [1,*], Simona Barison [4], Laura Fedele [2] and Sergio Bobbo [2]

1 Université Rennes 1, LGCGM, EA3913, F-35704 Rennes, France
2 Institute of Construction Technologies, National Research Council, I-35127 Padova, Italy
3 Departamento de Física Aplicada, Universidade de Vigo, E-36310 Vigo, Spain
4 Institute of Condensed Matter Chemistry and Technologies for Energy, National Research Council, I-35127 Padova, Italy
* Correspondence: dacabaleiro@uvigo.es (D.C.); patrice.estelle@univ-rennes1.fr (P.E.); Tel.: +33-(0)23234200 (P.E.)

† This article is an extended version of our paper published in 1st International Conference on Nanofluids (ICNf) and 2nd European Symposium on Nanofluids (ESNf), Castellón, Spain, 26–28 June 2019; pp. 118–121.

Received: 23 July 2019; Accepted: 24 August 2019; Published: 29 August 2019

Abstract: This work analyzes the dynamic viscosity, surface tension and wetting behavior of phase change material nano–emulsions (PCMEs) formulated at dispersed phase concentrations of 2, 4 and 10 wt.%. Paraffin–in–water samples were produced using a solvent–assisted route, starting from RT21HC technical grade paraffin with a nominal melting point at ~293–294 K. In order to evaluate the possible effect of paraffinic nucleating agents on those three properties, a nano–emulsion with 3.6% of RT21HC and 0.4% of RT55 (a paraffin wax with melting temperature at ~328 K) was also investigated. Dynamic viscosity strongly rose with increasing dispersed phase concentration, showing a maximum increase of 151% for the sample containing 10 wt.% of paraffin at 278 K. For that same nano–emulsion, a melting temperature of ~292.4 K and a recrystallization temperature of ~283.7 K (which agree with previous calorimetric results of that emulsion) were determined from rheological temperature sweeps. Nano–emulsions exhibited surface tensions considerably lower than those of water. Nevertheless, at some concentrations and temperatures, PCME values are slightly higher than surface tensions obtained for the corresponding water+SDS mixtures used to produce the nano–emulsions. This may be attributed to the fact that a portion of the surfactant is taking part of the interface between dispersed and continuous phase. Finally, although RT21HC–emulsions exhibited contact angles considerably inferior than those of distilled water, PCME sessile droplets did not rapidly spread as it happened for water+SDS with similar surfactant contents or for bulk–RT21HC.

Keywords: RT21HC paraffin; water; phase change material nano–emulsion (PCME); dynamic viscosity; surface tension; wetting behavior

---

1. Introduction

Phase change materials (PCMs) are substances that absorb/release large amounts of so–called "latent heat" when they undergo a phase change transition such as melting and freezing processes [1,2]. Thermal energy storage systems based on PCMs allow larger densities of energy storage within reduced temperature ranges, when compared to those reservoirs working with conventional thermal fluids (only based on sensible heat). Hence, phase change materials are considered potential energy saving materials to substantially improve the thermal energy storage capacity of a wide range of domestic or industrial facilities. However, an extended practical implementation of PCMs relies on their proper integration in thermal facilities. In this sense, phase change slurries have recently emerged as an interesting possibility [3]. PCM slurries are novel heat storage and transfer media that combine the latent capacity provided by the PCM and the good transport properties of the carrier fluid. Such slurries exhibit fluid–like behavior (they can be pumped) even when PCM droplets are solid or undergo phase transition. Therefore, these latent fluids may be directly integrated in thermal facilities without needing an additional thermal energy exchange [4]. Furthermore, PCM slurries allow higher heat transfer rates than only PCMs due to the large surface to volume ratio of the continuous phase [4].

Basically, there are three main types of phase changing slurries suitable for thermal management, namely hydrate slurries, micro–encapsulated/shape stabilized PCM suspensions and PCM–in–water emulsions [5]. Phase change material nano–emulsions (PCMEs) are formulated by directly dispersing fine droplets of a PCM into a carrier fluid with the assistance of appropriate surfactants. In this sense, PCMEs do not only avoid the supporting or coating materials necessary in micro–encapsulated/shape stabilized PCM suspensions, but they often exhibit lower rises in viscosity [4]. In addition, for phase change material emulsions, heat transfer is improved since there is a direct contact between the PCM and the carrier fluid (only separated by a thin surfactant layer) [6]. PCMEs have been investigated in the last years for different potential applications such as solar thermal facilities; heating, ventilation and air conditioning systems or increasing the thermal inertia of building components, among others [7,8]. An interesting approach to integrate PCM nano–emulsions is the use of capillary tube systems in cool ceiling and walls [6]. Thus, PCMEs with melting transitions on a temperature level of ~289–293 K can provide an additional density of storage energy in comparison to water normally utilized in such tubular mat systems. As an example, CryoSolplus20-in-water emulsions were developed by Huang et al. [9] for this kind of applications.

The evaluation of PCME thermal performance as storage and transfer media relies on an accurate characterization of the temperatures at which PCM nano–droplets undergo the phase change as well as the amounts of absorbed or released heat in those thermal events. However, the study of other thermophysical properties is also necessary to estimate the flow behavior and heat transfer performance of these materials. Thus, reliable viscosity data are essential for an appropriate selection and operation of equipment involved in formulation, storage or pumping of nano–emulsions. A revision of previous investigations on dynamic viscosity of PCM nano–emulsions or micro–encapsulated PCM suspensions shows that mass fraction of dispersed phase must usually be lower than 10–30 wt.% to ensure desired sample fluidity [5]. Inaba et al. [10,11] studied the dynamic viscosity of (*n*–tetradecane)–in–water emulsions at paraffin concentrations between 7.2 and 50.7 wt.%. Authors observed a pseudoplastic behavior and a slight change in viscosity when the PCM underwent liquid–solid phase change. Royon et al. [12] designed a paraffin–in–water emulsion containing a non–ionic surfactant and 50% of dispersed phase. In this case, dispersed phase droplets were ~2 μm in size and the melting point was 282.6 K. Studied sample showed a shear–thinning more pronounced when PCM droplets were solid. Also a slight jump in dynamic viscosity was observed during the solid–liquid transition of the dispersed phase.

Huang et al. [4,5] analyzed the rheological behavior of Rubitherm RT10–in–water emulsions prepared at paraffin loadings from 15 to 75 wt.% using a non–ionic surfactant (average PCM droplet sizes ~3–5 μm according to DLS measurements). In the flow curves (which cover the shear rate region up to 200 s^{-1}), samples were observed to exhibit a pseudoplastic behavior, which was more noticeable as the paraffin concentration increased. The addition of the PCM droplets also leaded to important rises in viscosity and consequently nano–emulsion viscosities were several times greater than that of water, even for the lowest concentration (15 wt.%). Authors found considerable higher viscosity increases at low temperatures (when PCM droplets are solid) and attributed this behavior to less deformation of solid particles. Wang et al. [13] produced water–based PCM emulsions using a mixture of polyvinyl alcohol and polyethylene glycol as emulsifier, 20 wt.% of paraffin (melting point ~331–333 K) as dispersed phase (droplet size 2–8 μm) and 0.05–0.1 wt.% of graphite as nucleating agent. Dynamic viscosity slightly decreased with the increasing shear rate (up to 100 s^{-1}). Agresti et al. [14]

recently studied the dynamic viscosity of aqueous nano–emulsions containing either commercial RT55 (melting temperature ~328 K) or RT70HC (~343 K) and stabilized with sodium dodecyl sulfate. Authors observed a strong non–Newtonian behavior in the case of RT70HC(10 wt.%)/water sample.

Some rheological studies on PCM nano–emulsions [4,5,9,13,15–17] or suspensions containing micro–encapsulated PCMs [18–21] also analyzed changes in dynamic viscosity occurring during the solidifyication/melting of phase change material droplets/micro–capsules. Most of those investigations observed deformations in shear viscosity–temperature evolution at temperatures around solid–liquid transitions. Huang and coworkers determined $\mu(T)$ curves through temperature ramp tests from 303 to 273 K and at cooling rates of 2 K/min for Rubitherm RT10–in–water emulsions [4,5] or acqueous PCMEs of CryoSol^{plus}6, CryoSol^{plus}10 and CryoSol^{plus}20 [9]. In those experiments, authors observed as emulsion viscosity rapidly increased once the dispersed paraffin began to solidify, especially at high PCM contents (>50% in mass). A similar change was found by Wang et al. [17] in heating shear rate–temperature sweeps performed in the temperature interval from 298 to 353 K for paraffin–in–water emulsions decorated with carbon nanomaterials. Such deformations in $\mu(T)$ curves were attributed to changes in the shape or volume of PCM droplets undergoing solid–liquid phase transition or to the creation to transient networks among neighboring droplets. In the case of micro–encapsulated paraffin slurries, Dutkowski and Fiuk [20] recently ascribed similar alterations in heating $\mu(T)$ sweeps to the chaotic movement of the non–molten core inside partially melted micro–capsules. Hence, non–uniformity in the solid/liquid state of paraffin droplets may increase internal friction resistance between carrier fluid layers.

Micro–structural rearrangements or spherical shape–restoration of emulsion droplets subject to shear stress can generate normal stresses [22]. Historically, changes in normal stresses were considered to the first evidence of elasticity in liquids [23]. Although elastic and loss moduli obtained from oscillatory tests are usually more appropriate to characterize changes in the structure of materials and these experiments have almost replaced shear normal stress measurements, these last analyses still provide useful information in certain situations. Thus, the study of first normal force difference (ΔN_1) has been found valuable in the characterization of polymers containing suspended particles [24,25], for example. When it comes to PCM slurries, an analysis of the temperature evolution of normal stress in shear flow could contribute to a better understanding of possible structural transformations occuring during solid–liquid phase change of PCM droplets/micro–capsules.

Nano–emulsions are kinetically–stable systems rather than thermodynamically–stable ones [26,27]. However, when emulsifier(s) and preparation conditions are appropriately selected, nano–emulsions can stay uniform for long periods of storage and so they are sometimes referred as "approaching thermodynamic stability" or "pseudo–stable" materials [28,29]. Despite main mechanisms involving emulsions stability being usually described by the "electric double layer model" and "DLVO" theory, surface tension is also considered as a possible factor influencing the stability of this kind of materials [26]. In addition, wetting behavior may also play a major role in the case of microfluidics systems such as the capillary tubes or micro–tubular configurations above mentioned. Thus, knowledge of surface tension is also required to understand the balance between viscous and surface tension competing forces that usually controls the actuation in micro–flows and is also likely to affect pressure–flow characteristics [30,31].

In this study, we analyzed the dynamic viscosity, surface tension and wetting behavior of RT21HC–in–water nano–emulsions prepared at 2, 4 and 10 wt.% dispersed phase concentrations. Experimental measurements were carried out at temperatures ranging from 275 to 303 K to evaluate whether the solid/liquid state of paraffin droplets had any effect on studied physical properties. With this work we intend to complete the characterization of this nano–emulsion system started in our recent publication [32], in which phase change transitions, thermal conductivities and volumetric behavior were analyzed.

2. Materials and Methods

2.1. Materials

Technical grade RT21HC paraffin ($T_{melt.}$= 293–294 K, $\Delta h_{melt.}$= 152 J/g) and RT55 wax ($T_{melt.}$= 328 K) were supplied by Rubitherm Technologies GmbH (Berlin, Germany). Water, W, (18.2 MΩ·cm at 298 K) was produced by a Millipore system (Billerica, MA, USA). Hexane (>98.5%) and sodium dodecyl sulfate, SDS, (98%) utilized in the emulsification process were purchased from Sigma Aldrich. A Sartorius analytical balance (Sartorius GmbH, Göttingen, Germany) with an uncertainty of 1×10^{-4} g was used to weigh reagents.

2.2. Nano–Emulsion Formulation

RT21HC–in–water PCMEs were formulated following the solvent–assisted method proposed in Agresti et al. [14]. First, two solutions, one of RT21HC in hexane (at a RT21HC: hexane mass ratio of 1:5) and the other of SDS in water, were joined at 318 K in a low–power ultrasonic bath. In a second step, obtained mixture was further sonicated using a VCX130 ultrasonic disruptor (Sonics & Materials, Inc., Newtown, CT, USA) working at 20 kHz and 65 W together with a 12 mm tip. The sample was then warmed up at 363 K and mechanically stirred for 2–3 h to completely evaporate the hexane. In this study, a surfactant:paraffin ratio of 1:8 was fixed to ensure fine emulsions with long temporal stabilities [14]. Thus, designed samples contained 2, 4 and 10 wt.% of RT21HC (dispersed phase) and 0.25, 0.5 and 1.25 wt.% of SDS (surfactant), respectively. An additional sample prepared using 3.6 wt.% of RT21HC and 0.4 wt.% of RT55 (nucleating agent) as dispersed phase and 0.5% of SDS was also studied for comparison.

The hydrodynamic sizes and ζ–potential values of paraffin droplets were determined at 298 K using a Zetasizer Nano–ZS90 (Malvern Instruments Ltd., Worcestershire, UK) based on Dynamic Light Scattering (DLS) method [33]. Figure 1a shows the size distributions obtained for the three emulsions designed without nucleating agent [32] just after PCME formulation.

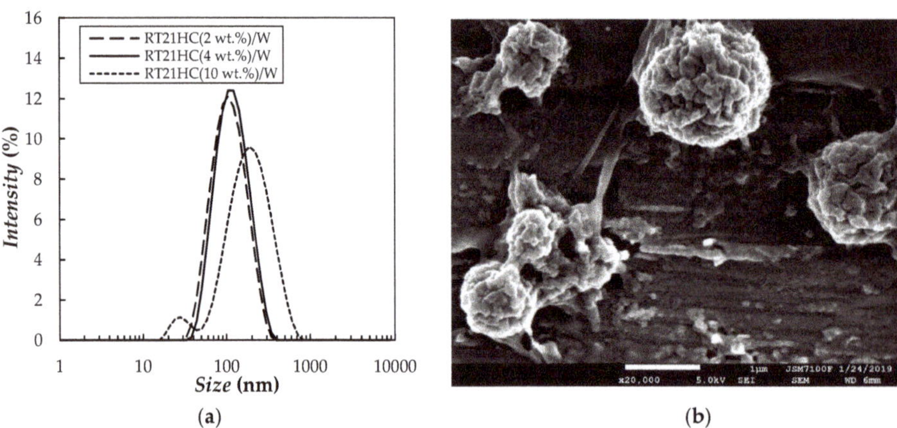

Figure 1. (**a**) Hydrodynamic droplet size distributions obtained by DLS [32] after PCME preparation and (**b**) SEM image of 4–month old RT21HC(2 wt.%)/water emulsion (1:20 diluted).

At 2 and 4 wt.% paraffin contents, PCM droplets exhibited average hydrodynamic sizes around ~90–100 nm and polydispersity indexes (*pdl*) ~0.17–0.23. A wider size distribution (*pdl*~ 0.28) was observed for RT21HC(10 wt.%)/water sample, with two main droplet populations around ~25 and ~180 nm. ζ–potential measurements showed negative values larger than −80 mV, indicating that electrostatically repulsive charges among droplets are strong enough to prevent agglomeration and coalescence. A thoughtful study of stability throughout storage, after freeze–thaw cycles and under mechanical shear was performed for these PCM nano–emulsions in our previous work [32].

The PCM droplet morphology of RT21HC(2 wt.%)/Water emulsion was also examined under a JSM-7100F thermal field scanning electron microscope (JEOL USA Inc., Peabody, MA, USA) operating at 5 kV. A four month–old sample was 1:20 diluted in water and sonicated in a Transonic TI–H–5 ultrasonic bath (Fisher Bioblock Scientific, Illkirch-Graffenstaden, France) operating at 550 W and 35 kHz for 5 minutes. The dilution was then deposited on the top of a carbon support and dried under atmospheric conditions. A representative SEM image (Figure 1b) shows that even agglomerated (due to the solvent evaporation necessary to carry out SEM studies), it is possible to intuit the sub–globular nano–metric morphology of the paraffin droplets.

2.3. Rheological Behavior

Shear viscosity (μ) was studied on a Malvern Kinexus Pro rheometer (Malvern Instruments Ltd., Worcestershire, U.K.) working with a cone–plate geometry (60 mm in diameter, a gap of 30 μm and 1° in cone angle). Temperature was controlled to within ±0.1 K by Peltier elements placed below the lower plate. Declared uncertainty of shear viscosity measurements with this device is better than 4% [34]. Two different types of experiments were carried out. First, shear viscosity–shear rate flow curves were obtained for water, water+SDS mixtures utilized to formulate the nano–emulsions, bulk–RT21HC and the four PCM–in–water emulsions. These tests were performed at 278.15 and 303.15 K except for the bulk paraffin (only studied in liquid phase, at 303.15 K). Viscosity values were collected in steady–state conditions at shear stresses logarithmically increasing from 0.008 to 3.5 Pa with at least 10 points per decimal, necessary to cover the range of shear rates between 10 and 1000 s^{-1} for all samples. In a second experiment, shear viscosity– and normal force–temperature curves in the range from 278 to 303 K were obtained at 0.2 and 1 K/min heating/scanning rates for water and RT21HC (10 wt.%)/water nano–emulsion.

2.4. Surface Properties

Interfacial tension (*IFT*), surface tension (*SFT*) and contact angle (*CA*) were studied by means of a drop shape analyser DSA–30 from Krüss GmbH (Hamburg, Germany) [35]. *IFT* and *SFT* determinations are based on pendant drop technique while *CA* investigations utilized sessile drop method. A TC40 environmental chamber (also from Krüss GmbH) was used during surface tension and contract angle analyses to control the temperature with a precision of ±0.2 K. Interfacial tension between bulk–RT21HC and water or between bulk–RT21HC and each of the three water+SDS mixtures used as base to produce the nano–emulsions were measured at room temperature (T = 297 ± 1 K). Liquid bulk–RT21HC was introduced in a glass container with ~40 ml of either water or water+SDS mixture using a J–shaped stainless needle (2.090 mm in gauge). Surface tensions at the air–sample surface were determined for distilled water, three water+SDS mixtures (at surfactant concentrations of 0.25, 0.5 and 1.25 wt.%), bulk–RT21HC and the four designed nano–emulsions using a 15–gauge needle (external diameter of 1.835 mm). Surface tension measurements were carried out in the range between 275 and 303 K, both increasing and decreasing temperature with steps of 2–5 K. The contact angles of water and nano–emulsion droplets formed using a 15–gauge needle and deposited on a stainless steel substrate were analyzed at 278 and 298 K using Young–Laplace equation. This substrate was previously utilized in the framework of a round robin test [36] in which nine independent European research centers of the Cost Action NanoUptake program [37] took part. A thoroughly characterization of substrate roughness is presented in Hernaiz et al. [36]. Recommendations such as substrate cleaning proposed in [36] were followed in this study. For the three physical properties, values were determined within a few seconds after drops were formed (*IFT* and *SFT*) or deposited in the stainless substrate (*CA*). Air–water interfacial tension was determined at room temperature and the result showed a deviation lower than 2% with surface tension data reported for water in [38]. Experimental uncertainty in measurements performed with this device was estimated to be 1% (*SFT*) [39], while a maximum relative deviation of 0.52% was obtained in [36] for this DSA–30 device measuring sessile drop gauges of 30°, 60° and 120° in CA.

3. Results and Discussion

3.1. Rheological Studies

3.1.1. Shear Viscosity–Shear Rate Flow Curves

Figure 2 shows the shear viscosity dependence on shear rate obtained at 278.15 and 303.15 K for water, bulk–RT21HC and four nano–emulsions. Flow curves of water+SDS mixtures prepared at surfactant concentrations of 0.25, 0.5 and 1.25 wt.% are also presented for comparison. Viscosity results measured for distilled water showed average deviations lower than 1.9% with previous literature [40]. A good agreement was also found between the value of 3.32 mPa·s determined in this work at 303.15 K for bulk–RT21HC and the previous data of 3.16 mPa·s reported at the same temperature for bulk–RT21 (another commercial paraffinic PCM with similar melting point) by Ferrer et al. [41].

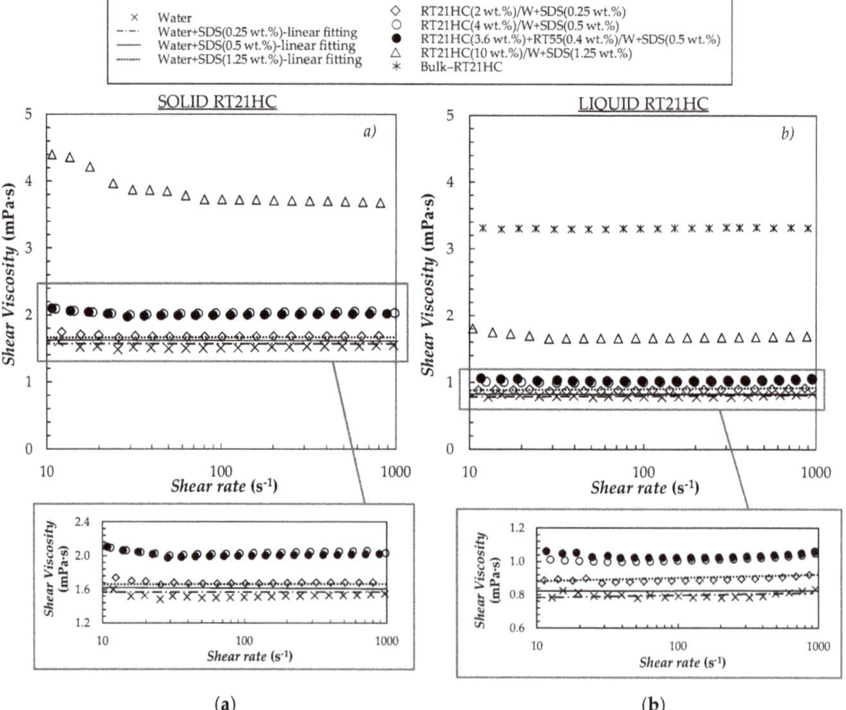

Figure 2. Shear viscosity–shear rate flow curves at (**a**) 278.15 K, when paraffin droplets are solid, and (**b**) 303.15 K, liquid paraffin droplets.

Like continuous and dispersed phases, nano–emulsions containing 2 and 4% mass concentrations of paraffin are mainly Newtonian within the studied shear rate range. However, a slight decrease in shear viscosity with increasing shear rate (pseudo–plastic behavior) was observed for the 10 wt.% paraffin loading in the region of low shear rates. As reviewed in the introduction, PCM nano–emulsions with shear–thinning behaviors were also reported in several studies [4,5,10–14]. Viscosity values and relative viscosities (defined as the ratio between sample and water viscosities at shear rates around ~100 s^{-1} (Newtonian region) are gathered in Table 1.

Table 1. Dynamic viscosities, μ, and viscosity increases regarding to water, μ/μ_{water}, obtained for the different samples at ~100 s^{-1}.

Sample	T = 278.15 K		T = 303.15 K	
	μ (mPa·s)	μ/μ_{water}	μ (mPa·s)	μ/μ_{water}
Water	1.50	1.00	0.791	1.000
Water+SDS(0.25 wt.%)	1.55	1.03	0.802	1.014
Water+SDS(0.5 wt.%)	1.61	1.07	0.813	1.028
Water+SDS(1.25 wt.%)	1.67	1.11	0.865	1.094
RT21HC(2 wt.%)/water+SDS(0.25 wt.%)	1.70	1.13	0.89	1.13
RT21HC(4 wt.%)/water+SDS(0.5 wt.%)	2.01	1.34	1.01	1.28
RT21HC(3.6 wt.%)+RT55(0.4 wt.%)/water+SDS(0.5 wt.%)	2.05	1.37	1.04	1.31
RT21HC(10 wt.%)/water+SDS(1.25 wt.%)	3.77	2.51	1.67	2.11
Bulk–RT21HC	-	-	3.32	-

As can be observed, viscosity strongly rises with PCM+SDS loading, reaching increases up to 151% in the case of RT21HC(10 wt.%)/Water nano–emulsion. Larger viscosity increments were observed at 278.15 K (when the material of droplets is solid) than at 303.15 K (liquid droplets). This is more appreciable in the case of RT21HC(10 wt.%)/water sample, for which indicated difference reached 39%. A similar behavior was reported by Mikkola et al. [42] when the viscosity of studied paraffin–in–water emulsions using either a falling ball viscosimeter or a rotational rheometer. In that last investigation, authors reported μ/μ_{water} ratios of 2.35 (at temperatures at which paraffin droplets were solid) and 1.95 (liquid paraffin droplets) for a dispersed phase concentration of 10 wt.%. Those viscosity ratios are close in value to the results presented here for the same PCM load.

Oil–in–water emulsions with micro– or macro–metric droplets are known to usually exhibit lower viscosity ratios (regarding base fluid) than solid–water suspensions with the same particle size. This fact can be attributed to three different causes. (i) When dispersed phase is liquid, material inside droplets tends to roll around in a "tank–tread–like" motion during sample flow [23]. The friction caused by relative movements between layers of dispersed material makes that the moment of rotation transmitted to the surface layers gets reduced as it moves towards the center of the drops. This internal circulation in liquid particles reduces resulting droplet rotation and, consequently, flow disturbance [20]. The higher the dispersed phase to continuous phase viscosity ratio (i.e., $z = \mu_{PCM}/\mu_{water}$), the more deformed will get the streamlines (approaching the limit of droplet viscosity of rigid–sphere suspensions). However, given the nano–metric size of PCM droplets present in nano–emulsions, this phenomenon should have a much weaker impact than in the case of micro– or macro–emulsions. (ii) Liquid droplets are more easily deformable under a shear stress field. The influence of deformation rate on suspensions viscosity depends on μ_{PCM}/μ_{water} as well as on drop relaxation time, λ_d, which can be calculated as follows [23]:

$$\lambda_d = \frac{r \cdot u_{water}}{\Gamma} \quad (1)$$

where r is the radius of the PCM droplet, μ_{water} is the viscosity of the surrounding media and Γ is the interfacial tension between droplets and base fluid. In our case, even in the nano–emulsion containing 10 wt.% of RT21HC, individual paraffin droplets exhibited average hydrodynamic sizes lower than ~160 nm (value measured 1 month after preparation) [32], while interfacial tension between RT21HC and water was ~0.042 N/m (Γ reduced to ~0.008 N/m when SDS was added to water, see Section 3.2.1). Thus, droplet relaxation times are expected to be very short, $\lambda_d < 2 \times 10^{-9}$ s, and drop deformation very small, even for relatively high shear rates [23]. iii) Elastic or non–elastic nature of collisions among neighboring PCM droplets, which will also depend on $z = \mu_{PCM}/\mu_{water}$, or even among possible clusters of (solvated/swollen) droplets. We consider this last explanation as the most feasible, although more research is needed to deep in the complex behavior of this kind of nano–metric structures and confirm this fact.

Numerous attempts have been carried out in the literature to develop theoretical or empirical equations to describe the shear viscosity of emulsions or solid particle dispersions as a function of dispersed phase concentration [43–45]. The relative viscosity ($\mu_r = \mu/\mu_0$), which represents the ratio between dispersion viscosity (μ) and base fluid viscosity (we will consider that of corresponding water+SDS mixture used to produce the sample, $\mu_0 = \mu_{water+SDS}$), can be formally written as a virial of series:

$$u_r \approx 1 + |\eta| \cdot \phi + k_H \cdot \phi^2 + \ldots \quad (2)$$

where $|\eta|$ is the intrinsic viscosity and k_H is known as Huggins' coefficient. Intrinsic viscosity is usually assumed to be $|\eta|$ = 5/2 for dispersions of hard spheres, while for spherical liquid droplets it takes a value $1 \leq |\eta| \leq 5/2$, which can be calculated as follows [46]:

$$|\eta| = 1 + \frac{3 \cdot z/2}{1+z} \quad (3)$$

where z the dispersed phase to continuous phase viscosity ratio (i.e., $z = \mu_{PCM}/\mu_{water}$). In the case of non–interacting particles, Equation (2) can be reduced to the linear equation, $\mu_r \approx 1 + |\eta| \cdot \phi$, first proposed by Einstein [47]. However, for dispersions with $\phi > 0.02$ (in the case of spherical or non–interacting particles) viscosity increases more rapidly with concentration than predicted by that linear approximation and so other terms of Equation (2) must be considered.

When concentration overcomes the limit known as geometrical percolation threshold (p_c), rises in viscosity with increasing particle load are usually more pronounced than those predicted by Einstein's linear approximation or virial expansions of viscosity based on Equation (2) [27,44]. Beyond percolation concentration, samples start forming large contracting clusters consisting of a flocculated structure in which primary particles and base fluid are trapped [44]. Thus, although sample continues being fluid–like (the transition from viscous fluid to true solid happens at the concentration known as maximum packing volume fraction, ϕ_m, which is usually $\phi_m >> p_c$) motions are necessarily collective [48]. Thus, increases in viscosity can correspond to those values expected according to theoretical or semi–empirical equations at higher concentrations. In order to take into account the effect of solvation and/or aggregation of droplets, an effective volume fraction of dispersed phase can be defined as:

$$\phi_{eff} = k_a \cdot \phi_s \quad (4)$$

k_a is the aggregation coefficient and ϕ_s is the volume fraction of solvated or swollen droplets. Solvated volume fraction can be calculated as [27,49]:

$$\phi_s = \phi \cdot (1 + \delta/r)^3 \quad (5)$$

where δ and r are the thickness of the solvated layer and the radius of individual nano–droplets, respectively. The aggregation coefficient usually depends on ϕ_s. Different equations have been proposed to calculate k_a [49], in this work we have considered the three following expressions:

$$k_a = \frac{\phi_{eff}}{\phi_s} = \frac{1}{\phi_m} \quad (6)$$

$$k_a = \frac{\phi_{eff}}{\phi_S} = 1 + \phi_S \cdot \left(\frac{1-\phi_m}{\phi_m}\right) \cdot \left[\sqrt{1 - \left(\frac{\phi_m - \phi_S}{\phi_m}\right)^2}\right] \quad (7)$$

$$k_a = \frac{\phi_{eff}}{\phi_S} = \frac{1}{\phi_S} \cdot \left[1 - \exp\left(-\frac{\phi_S}{1 - \left(\frac{\phi_S}{\phi_m}\right)}\right)\right] \quad (8)$$

Equation (6) assumes that k_a coefficient is independent of the volume fraction of solvated droplets. Equation (7) was developed by Pal et al. [50] considering the constraints that $k \rightarrow 1$ when $\phi_s \rightarrow 0$ and $dk/d\phi_s \rightarrow 1$ when $\phi_s \rightarrow \phi_m$. Equation (8) was considered in [51,52] for large micron–sized suspensions.

One of the most widely used models to predict the viscosity of emulsions or solid–liquid suspensions is that presented by Krieger and Dougherty (K–D) [53]. Derived by differential effective medium theory from a previous viscosity–concentration relationship proposed by Brinkman [54] and Roscoe [55], the K–D equation can be expressed as:

$$\mu_r = \left(1 - \frac{\phi}{\phi_m}\right)^{-|\eta|\cdot\phi_m} \tag{9}$$

In order to consider the effects of droplet solvation and aggregation in Krieger and Dougherty model, ϕ_{eff} was used instead of ϕ in Equation (9). Maximum packing volume fraction was assumed to be ϕ_m= 0.637 (spherical shape), while the thickness of the solvated layer was considered as a fitting parameter as proposed by Pal [49].

In the case of nano–emulsions stabilized with ionic surfactants (such as the SDS used in this work), effective/relative dynamic viscosity was observed to increase with dispersed phase volume fraction much more rapidly than for emulsions of larger droplets [27,56]. Figure 3 shows the effective viscosities of the studied PCMEs together with the values calculated with Einstein [47] and modified Krieger and Dougherty [53] models. It must be pointed out that a proper comparison between different nano–emulsions would require that all samples would contain the same amount of surfactant.

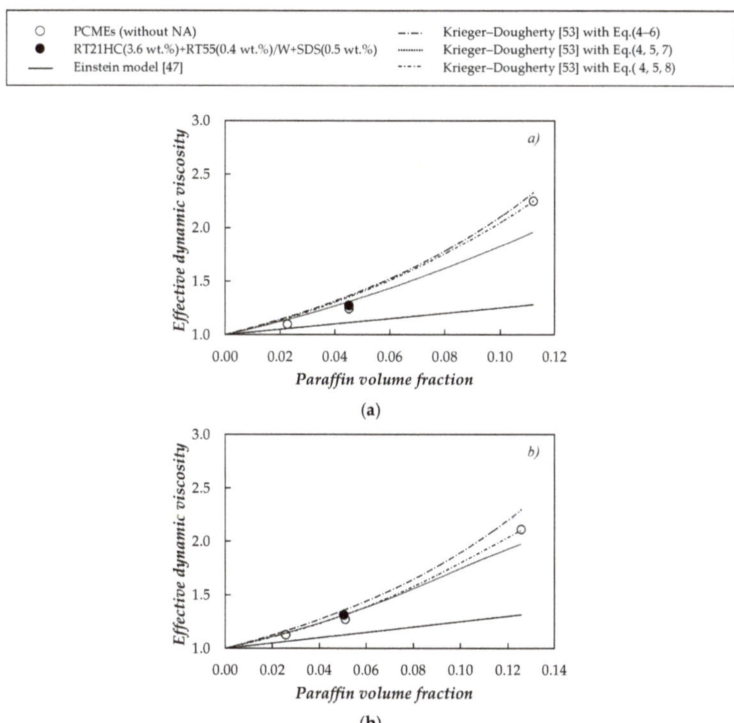

Figure 3. Volume fraction, ϕ, dependence of effective dynamic viscosity, $\mu_r = \mu/\mu_{\text{water+SDS}}$, at (**a**) 278.15 K and (**b**) 303.15 K.

As can be observed, μ_r exhibits a rather non–linear dependence on volumetric fraction; something expected since $\phi > 0.02$ for all studied samples. Neither the introduction of $k_H \times \phi^2$ improves the predictions of μ_r (not graphically presented in Figure 3. Better results (*AADs* < 6%) were obtained combining the KD [53] equation with equations (4–8) to consider the effect of droplet solvation and aggregation. In particular, equations (4, 5, 8, 9) combination described experimental results at 278.15 K (solid droplets with an *AADs%* of 4.2% (considering a solvated layer of $\delta \sim 7.5$ nm), while this value reduced to 1.2% ($\delta \sim 4.8$ nm) at 303.15 K (liquid droplets). Similar thicknesses of solvated layers were estimated by Pal [49] for nano–emulsions ($\delta \sim 2.6$–5.6 nm) with droplet diameters in the range of 28–205 nm or solid–liquid suspensions ($\delta \sim 3.0$–13.8 nm) with nanoparticle sizes of 29–146 nm using a similar approach.

3.1.2. Shear Viscosity–Temperature Sweeps

With the objective of analyzing PCMEs viscosity during droplets phase change, temperature ramp tests were performed at constant shear stress for water as well as 4 wt.% and 10 wt.% RT21HC/Water nano–emulsions. Samples were heated/cooled between 278 and 303 K at scanning rates of 0.2 and 1 K/min. A shear stress of 0.4 Pa, which corresponds to shear rates ranging from ~100 s^{-1} (at 278 K) to 240 s^{-1} (at 303 K) in the case of RT21HC(10 wt.%)/Water, was applied to ensure that viscosity values were collected in the Newtonian region (see Figure 2). The temperature evolutions of viscosity for water and the two analyzed emulsions at the two scanning rates are shown in Figure 4.

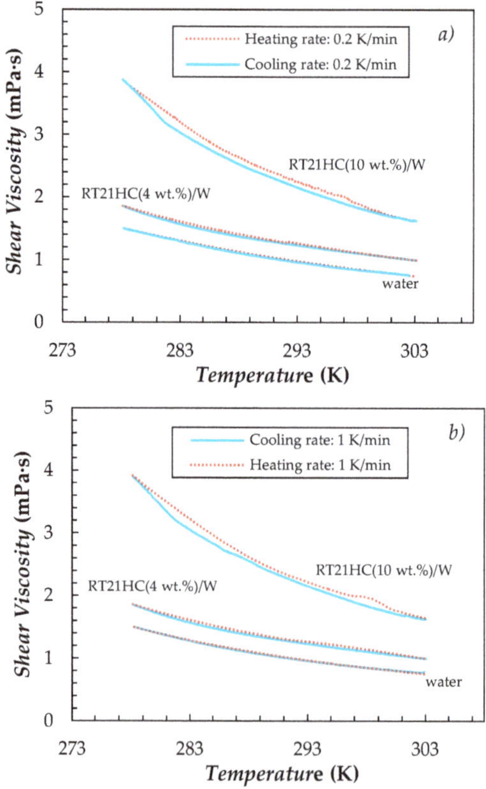

Figure 4. Shear viscosity–temperature curves obtained for water as well as 4 wt.% and 10 wt.% RT21HC/Water emulsions at scanning rates of (**a**) 0.2 and (**b**) 1 K/min.

Following the universal behavior of liquids, dynamic viscosity decreases with increasing temperature. At 278 and 303 K (extremes of temperature ramps), μ values determined from flow curve experiments and temperature sweeps are in good agreement. For RT21HC(4 wt.%)/water sample, differences in cooling and heating $\mu(T)$ sweeps are within experimental uncertainty. However, in the case of RT21HC(10 wt.%)/water sample, viscosities obtained in the range from 283 to 291 K are higher when paraffin droplets are solid (heating curve) than when they are liquid (cooling ramp). Minimum differences between the two μ values (for RT21HC(10 wt.%)/water in the mentioned temperature interval) are ~4–6%. In addition, and as previously reported in the literature for other PCM–in–water slurries [13,15–17,57], shear viscosity–temperature curves are affected by melting and recrystallization processes. Such deformations are only appreciated in $\mu(T)$ curves of the nano–emulsion with 10 wt.% of paraffin (due to higher content in droplets undergoing the phase change). Thus, as temperature increases in heating curves, a change in the exponential decreasing trend of viscosity of RT21HC(10 wt.%)/water sample is observed at temperatures ~292–300 K. A similar alteration in $\mu(T)$ curves is detected at temperatures ~278–283 K in cooling ramps. During the melting process of paraffin droplets (onset temperature melting transition is ~291.5 K according to [32]), an additional amount of energy is required so that the RT21HC contained in the sample undergoes the solid–liquid transition. This supplementary energy can be responsible (at least in part) for that deformation in $\mu(T)$ curve, since the temperature of the carrier fluid closer to paraffin droplets may rise at a slower pace than it does before the melting transition of dispersed phase. As a consequence, higher apparent viscosities are measured. A similar description could be used to explain the lower viscosities observed in the cooling sweeps. As reviewed in the introduction, other authors have also attributed these "peaks" in viscosity–temperature curves to changes in the shape or volume of PCM droplets undergoing solid–liquid phase transition. In our case, slight differences (within experimental uncertainty) were observed between the size distributions obtained for these same nano–emulsions in the temperature range from 276 K (solid paraffin droplets) to 303 K (liquid droplets) by DLS [32]. However, paraffin droplets are expected to expand when passing from solid to liquid (due to change in density of bulk–RT21HC). In fact, a volumetric study of this nano–emulsion system [32] showed as density of RT21HC(10 wt.%)/Water sample reduced by 1.1% when heated from 292 to 296 K, while pure water only exhibited a density reduction of 0.1% in the same temperature range. Another plausible explanation was recently proposed by Dutkowski and Fiuk [20,21] when the studied Micronal®DS 2039 X, an aqueous slurry containing paraffin micro–capsules coated with a highly cross–linked polymethyl methacrylate. Authors attributed similar deformations in heating $\mu(T)$ curves around phase change to the chaotic movement of the non–molten core inside partially melted micro–capsules. Thus, this non–uniformity in the solid/liquid state of droplets was assumed to be responsible for an increase in internal friction resistance between carrier fluid layers.

In order to further study possible structural rearrangements occuring during solid–liquid phase change transitions of PCM droplets, changes in normal stress during shear viscosity-temperature experiments were also analyzed. The temperature evolutions of first normal force difference for RT21HC(10 wt.%)/Water sample at the two scanning rates are shown in Figure 5. Heating and cooling thermograms at a scanning rate of 1 K/min obtained by differential scanning calorimetry in [32] are also shown for comparison (Figure 5c).

In our case, negative first normal force differences were observed. This is due to that, during the rotation of a Newtonian fluid without strong interactions among suspended particles, inertia forces cause sample to flow away from the center [22]. Experiments were performed at constant shear stress. Thus, as temperature rises, sample viscosity decreases allowing cone geometry to turn faster (shear rate increases) and, in turn, ΔN_1 becomes more negative. Likewise, negative ΔN_1 reduces (in absolute value) as temperature decreases. Two different changes are observed in heating and cooling curves around melting and recrystallization temperatures, respectively. In heating process there is a change in the slope of the curve at ~292 K. This value agrees quite well with the melting onset temperature of 291.5 K obtained in [32] for this same sample by Differential Scanning Calorimetry using scanning rates of 1 K/min (see Figure 5c). This slight change in ΔN_1–trend may be due to the reduction in sample viscosity as consequence of the solid to liquid transition of paraffin droplets above described. A stronger change was observed at temperatures ~283 K, which agrees with the crystallization onset temperature of 282.2 K [32]. This more remarkable variation may be attributed to the super–cooling exhibited by the nano–emulsions. Thus, it is well reported that, after crystallization begins, temperature tends to rise and approach melting point [58].

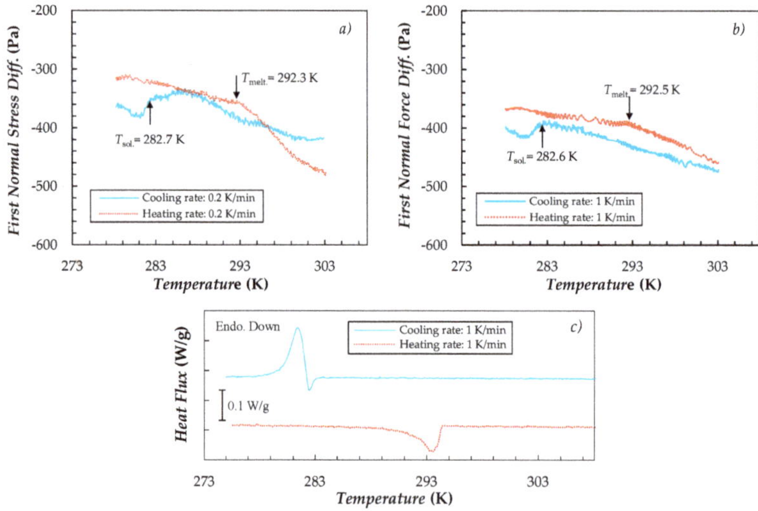

Figure 5. (**a**,**b**) First normal stress difference– temperature curves obtained for RT21HC(10 wt.%)/Water emulsion at scanning rates of (**a**) 0.2 and (**b**) 1 K/min. (**c**) Thermogram obtained by differential scanning calorimetry in [32].

3.2. Surface Properties

3.2.1. Interfacial Tension

Interfacial tensions (*IFT*) experimentally obtained at room temperature are gathered in Table 2. The *ITF* of 41.95 mN/m measured between bulk–RT21HC and water in this work was lower than the range of ~50–55 mN/m usually reported at 298 K for different alkane–water systems elsewhere [59]. An *ITF*< 42 mN/m was also obtained at 295 K between *n*–decane and water by Goebel and Lunkenheimer [60] when using the alkane as received (without any purification). However, the same authors found that *ITF* increased up to 53 mN/m after *n*–decane was purified. Those differences in *ITF* were attributed to unwanted amphiphilic residues from the oxidation process used during alkane separation and fractionation [60].

Table 2. Interfacial tensions, *ITF*, between bulk–RT21HC and water or between bulk–RT21HC and the three water+SDS mixtures experimentally measured at room temperature.

System	*T* (K)	*ITF* (mN/m)
RT21HC–Water	297.06 ± 0.04	41.95 ± 0.34
RT21HC–Water+SDS(0.25 wt.%)	296.57 ± 0.07	8.32 ± 0.09
RT21HC–Water+SDS(0.5 wt.%)	297.21 ± 0.03	8.20 ± 0.08
RT21HC–Water+SDS(1.25 wt.%)	297.44 ± 0.02	8.19 ± 0.14

As it can be observed in Table 2, *ITF* between paraffin and water considerably reduced with the addition of SDS to the water. Our results ~8.2–8.3 mN/m were also lower than the *ITF* of 9.65 mN/m (298 K) reported by Oh and Shah [61] for *IFT* between *n*–hexadecane and water+SDS mixtures at surfactant concentrations larger than 8 mM (0.25 wt.% surfactant concentration studied in this work corresponds to ~8.6 mM).

3.2.2. Surface Tension

Surface tensions (*SFTs*) experimentally measured for water, the water+SDS mixture containing 0.25 wt.% of surfactant, bulk–RT21HC (in liquid phase) and the four PCMEs are plotted in Figure 6.

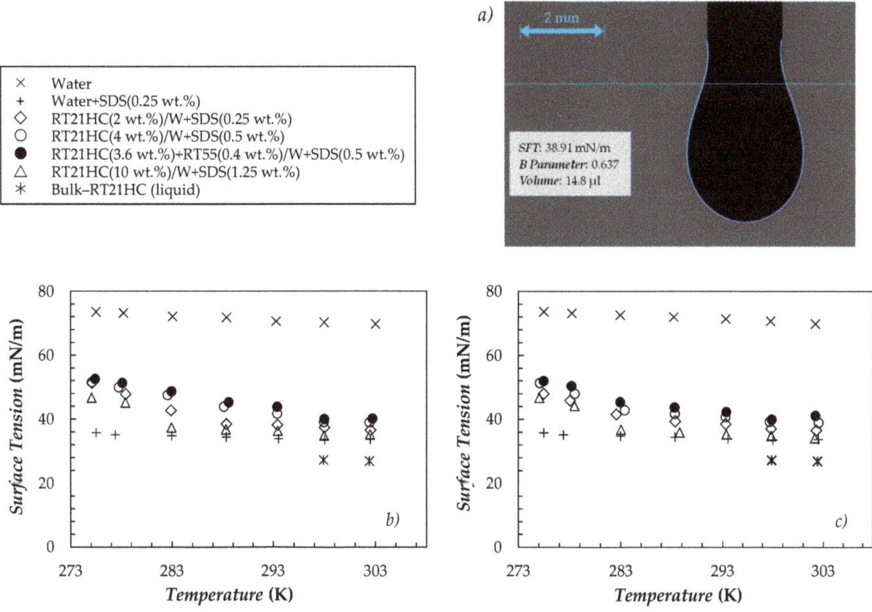

Figure 6. (**a**) Droplet capture obtained at 303 K for the RT21HC(4 wt.%)/Water nano–emulsion. (**b**,**c**) Surface tension results of the different samples obtained from (**b**) increasing and (**c**) decreasing temperature experiments.

Water results showed an *AAD*% of 2.3% with previous surface tension data reported in [38]. As expected, the addition of SDS considerably reduced the surface tension of water. In our case, similar results were measured for the water+SDS mixtures at surfactant concentrations of 0.25, 0.5 and 1.25 wt.% (maximum deviations among the three systems were found at temperatures ~278 K and did not exceeded 3%). The critical micelle concentration of water+SDS system has been previously reported in literature to be around ~8 mM (corresponding to ~0.2–0.25 wt.%) [62] although certain authors reduced it to ~5 mM [26]. Bulk–RT21HC (in liquid phase) exhibited *SFTs* of 26.8–27.2 mN/m, which agrees well with the value of 27.2 mN/m reported by Goebel and Lunkenheimer [60] for *n*–hexadecane (an alkane with melting point at ~291 K).

Important reductions in surface tension (in comparison to water) were also observed for the PCMEs. Nevertheless, the *SFTs* of nano–emulsions were higher than the values obtained for the water+SDS mixtures with the same surfactant concentration. A similar behavior was found by Zhang et al. [26] when studied poly(butyl acrylate)–in–water emulsions at dispersed phase concentrations between 7 and 22% and droplets sizes in the range of 100–500 nm. In that study, authors observed that the minimum *SFT* value of ~31–33 mN/m obtained for water+SDS with surfactant loadings larger than ~5 mM (CMC), was not reached up to SDS concentrations of 60–80 mM in the case of an emulsion containing 22 wt.% of poly(butyl acrylate) droplets with average size of 115 nm. This was attributed to the fact that part of the surfactant is evenly absorbed by the droplets and also that dispersed phase droplets introduce additional surface to be covered by the surfactant. As a consequence of that, Zhang et al. [26] found that, for the same SDS concentration and before CMC was reached, *SFT* increased with increasing dispersed phase concentration and with reducing dispersed phase droplet size. This behavior agrees with what can be observed in Figure 6. Thus, despite RT21HC(4 wt.%)/water and RT21HC(3.6 wt.%)+RT55(0.4 wt.%)/water samples containing twice as much SDS as RT21HC(2 wt.%)/water emulsions, the two first samples exhibit slightly higher *SFT*s in comparison with 2 wt.% concentration. 2 and 4 wt.% emulsions have similar average paraffin droplet sizes (see Figure 1a). *SFT* measurements were performed at increasing and decreasing temperatures to analyze whether the solid/liquid state of paraffin droplets could have any influence on this physical property. Differences in *SFT* between PCMEs and water+SDS mixtures are higher at lower temperatures (when paraffin droplets are solid). In our opinion, such behavior may be due to the fact that during the solidification of the drops, part of the surfactant remains anchored or completely trapped by the paraffin. As a consequence, an additional surface is available to be covered by surfactant molecules.

3.2.3. Wetting Behavior

Contact angle (*CA*) measurements were carried out for distilled water and the four nano–emulsions at two characteristic temperatures 278 and 298 K at which paraffin dispersed phase is solid and liquid, respectively. Analyzed PCME volumes were ~11–14 μL while they were ~24 μL (278 K) and 21 μL (298 K) in the case of water. Figure 7 shows some examples of sessile drops, while *CA* results are gathered in Table 3.

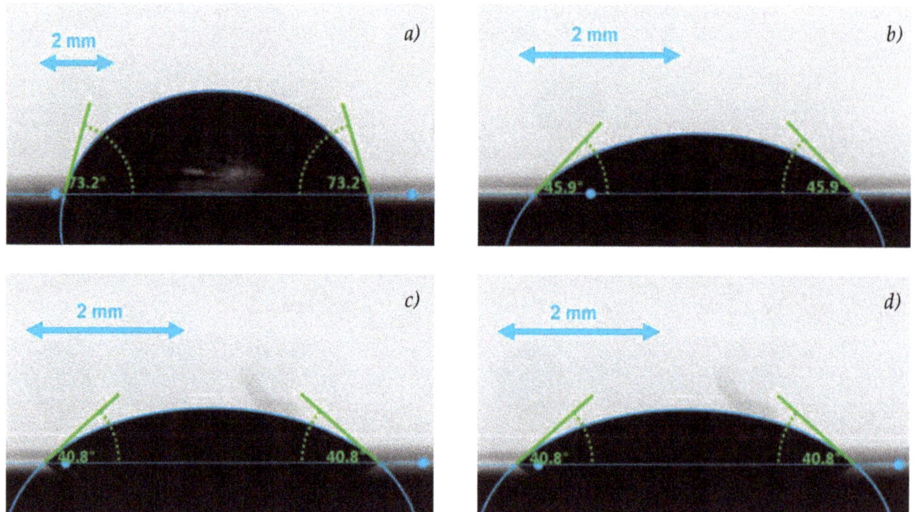

Figure 7. Examples of sessile drops analyzed at 278 K for: (**a**) water; (**b**) RT21HC(2 wt.%)/Water; (**c**) RT21HC(4 wt.%)/Water and (**d**) RT21HC(10 wt.%)/Water.

Table 3. Contact angles, *CA*, measured for distilled water and studied nano–emulsions on a stainless steel substrate [36]. Relative humidity: 50 ± 10%.

Sample	*CA* (°)	
	***T* = 278 K**	***T* = 298 K**
Water	72 ± 2	69 ± 2
RT21HC(2 wt.%)/W+SDS(0.25 wt.%)	46 ± 2	37 ± 2
RT21HC(4 wt.%)/W+SDS(0.5 wt.%)	41 ± 2	35 ± 2
RT21HC(3.6 wt.%)+RT55(0.4 wt.%)/W+SDS(0.5 wt.%)	41 ± 1	36 ± 2
RT21HC(10 wt.%)/W+SDS(1.25 wt.%)	40 ± 2	33 ± 2

Water *CA* at 298 K is close in value to those CAs reported for sessile drops of similar volumes in Hernaiz et al. [36]. Unlike the water+SDS mixtures at surfactant concentrations of 0.25, 0.5 and 1.25 wt.% or bulk–RT21HC (whose *CAs* could not be measured since samples rapidly spread on stainless steel substrate), RT21HC–in–water emulsions exhibited *CAs* of 40–46° (278 K) and 33–37° (298 K). Although PCME values are up to 52% lower than those of water, it is clear that the presence of possible free surfactant in the emulsions did not overcome the cohesive forces in the RT21HC–SDS+water system. A comparison between the different nano–emulsions seems to show a slight decrease in *CA* with increasing paraffin concentration. However, differences in *CA* between most of the samples are within standard deviations. In this case, a stainless steel solid substrate [36] was selected as representative surface in industrial applications. Other substrates may be interesting depending on the particular use of the nano–emulsions.

4. Conclusions

Phase change material nano–emulsions produced following a solvent–assisted route were characterized in terms of dynamic viscosity and surface properties. A slight shear–thinning behavior was observed in the region of low shear rates (up to 30–60 s^{-1}) for the nano–emulsion prepared at the highest concentration (10 wt.%), while the rest of studied samples exhibited a Newtonian behavior at studied conditions. As expected, dynamic viscosity strongly rose with dispersed phase+surfactant concentration. Within the Newtonian region, maximum μ increases (regarding distilled water) reached 151% for the paraffin concentration of 10 wt.% at 278 K. Larger increases in viscosity were observed when paraffin droplets are solid than when they are liquid. A good description (with deviations lower than 6%) was obtained when Krieger–Dougherty $\mu(\phi)$ model was modified to include solvated/swollen and aggregation particle effects. Solidification and recrystallization temperatures determined from rheological temperature–sweeps for RT21HC(10 wt.%)/Water emulsion agreed with previous calorimetric analysis using DSC. Surface tension reduces up to 50% with the presence of paraffin droplets. However, these diminutions are lower than the decreases found for the corresponding water+SDS mixtures used to produce the nano–emulsions. PCME contact angles are up to 52% lower than those of water, but sample did not rapidly spread in stainless steel substrate as it happened to bulk–RT21HC or SDS+water mixtures without paraffin droplets. No significant effect on any of the three studied physical properties was observed when part of the dispersed phase was replaced by a nucleating agent.

Author Contributions: F.A. and S.B. (Simona Barison) defined PCME formulation, prepared studied samples and helped in data interpretation and writing. D.C. and S.H. measured rheological and surface properties and first drafted the manuscript. P.E., L.F. and S.B. (Sergio Bobbo) conceived the study, planned the experiments and took an active role in the preparation of the manuscript. All authors read and approved the final manuscript.

Funding: D.C.: S.H., F.A., S.B. (Simona Barison) and S.B. (Sergio Bobbo) acknowledge the EU COST Action CA15119: Overcoming Barriers to Nanofluids Market Uptake for financial support in the participation of the 1st International Conference on Nanofluids (ICNf) and the 2nd European Symposium on Nanofluids (ESNf) held at the University of Castellón, Spain during 26–28 June 2019. P.E. acknowledges the European Union through the European Regional Development Fund (ERDF), the Ministry of Higher Education and Research, the French region of Brittany and Rennes Métropole for the financial support related to the device used in this study for surface tension and contact angle measurements. D.C. is recipient of a postdoctoral fellowship from Xunta de Galicia (Spain).

Acknowledgments: This investigation is a contribution to the COST (European Cooperation in Science and Technology) Action CA15119: Overcoming Barriers to Nanofluids Market Uptake (NanoUptake). Authors acknowledge Francis Gouttefangeas from CMEBA (Université Rennes 1) for performing SEM studies.

Conflicts of Interest: The authors declare no conflict of interest.

Nomenclature

$AAD\%$	percentage absolute average deviations
CA	contact angle [°]
CMC	critical micelle concentration
ΔN_1	first normal force difference [Pa]
$\lvert\eta\rvert$	intrinsic viscosity
ϕ	volume fraction
ϕ_{eff}	effective volume fraction
ϕ_m	maximum packing volume fraction
ϕ_s	solvated/swollen droplet volume fraction
ITF, Γ	interfacial tension [mN/m]
k_a	aggregation coefficient
K–D	Krieger and Dougherty viscosity model
k_H	Huggins coefficient
p_c	geometrical percolation threshold
PCM	phase change material
PCME	phase change material emulsion
pdI	polydispersity index
r	radius of PCM droplet [m]
RT21HC	Rubitherm RT21HC paraffin
RT55	Rubitherm RT55 paraffin wax
SDS	sodium dodecyl sulfate
SFT	surface tension [mN/m]
T	temperature [K]
W	water
z	dispersed phase to continuous phase viscosity ratio ($z = \mu_{PCM}/\mu_{water}$)
δ	thickness of solvated layer
λ_d	drop relaxation time
μ	shear dynamic viscosity [mPa·s]
μ_r	relative viscosity of sample to corresponding water+SDS mixture ($\mu_r = \mu_{PCM}/\mu_{water+SDS}$)
Subscripts	
melt.	melting
PCM	phase change materials
PCME	phase change material emulsion
sol.	solidification

References

1. Singh, S.; Gaikwad, K.K.; Suk, Y. Phase change materials for advanced cooling packaging. *Environ. Chem. Lett.* **2018**, *16*, 845–859. [CrossRef]
2. Cabeza, L.F.; Castell, A.; Barreneche, C.; De Gracia, A.; Fernández, A.I. Materials used as PCM in thermal energy storage in buildings: A review. *Renew. Sustain. Energy Rev.* **2011**, *15*, 1675–1695. [CrossRef]
3. Zhang, P.; Ma, Z.W.; Wang, R.Z. An overview of phase change material slurries: MPCS and CHS. *Renew. Sustain. Energy Rev.* **2010**, *14*, 598–614. [CrossRef]
4. Huang, L.; Petermann, M.; Doetsch, C. Evaluation of paraffin/water emulsion as a phase change slurry for cooling applications. *Energy* **2009**, *34*, 1145–1155. [CrossRef]
5. Huang, L.; Petermann, M. An experimental study on rheological behaviors of paraffin/water phase change emulsion. *Int. J. Heat Mass Transf.* **2015**, *83*, 479–486. [CrossRef]
6. Youssef, Z.; Delahaye, A.; Huang, L.; Trinquet, F.; Fournaison, L.; Pollerberg, C.; Doetsch, C. State of the art on phase change material slurries. *Energy Convers. Manag.* **2013**, *65*, 120–132. [CrossRef]
7. Tesfai, W.; Singh, P.; Shatilla, Y.; Iqbal, M.Z.; Abdala, A.A. Rheology and microstructure of dilute graphene oxide suspension. *J. Nanoparticle Res.* **2013**, *15*, 1989. [CrossRef]

8. Delgado, M.; Lázaro, A.; Mazo, J.; Zalba, B. Review on phase change material emulsions and microencapsulated phase change material slurries: Materials, heat transfer studies and applications. *Renew. Sustain. Energy Rev.* **2012**, *16*, 253–273. [CrossRef]
9. Huang, L.; Doetsch, C.; Pollerberg, C. Low temperature paraffin phase change emulsions. *Int. J. Refrig.* **2010**, *33*, 1583–1589. [CrossRef]
10. Inaba, H.; Morita, S.; Nozu, S. Fundamental study of cold heat-storage system of phase-change-type emulsion having cold latent heat dispersion material. *Trans. JSME* **1993**, *59*, 2882–2889. [CrossRef]
11. Inaba, H.; Morita, S. Flow and cold heat-storage characteristics of phase-change emulsion in a coiled double-tube heat exchanger. *J. Heat Transf.* **2016**, *117*, 440–446. [CrossRef]
12. Royon, L. Physical properties and thermorheological behaviour of a dispersio having cold latent heat-storage material. *Energy Convers. Manag.* **1998**, *39*, 1529–1535. [CrossRef]
13. Wang, F.; Liu, J.; Fang, X.; Zhang, Z. Graphite nanoparticles-dispersed paraffin/water emulsion with enhanced thermal-physical property and photo-thermal performance. *Sol. Energy Mater. Sol. Cells* **2016**, *147*, 101–107. [CrossRef]
14. Agresti, F.; Fedele, L.; Rossi, S.; Cabaleiro, D.; Bobbo, S.; Ischia, G.; Barison, S. Nano-encapsulated PCM emulsions prepared by a solvent-assisted method for solar applications. *Sol. Energy Mater Sol. Cells* **2019**, *194*, 268–275. [CrossRef]
15. Lu, W.; Tassou, S.A. Experimental study of the thermal characteristics of phase change slurries for active cooling. *Appl. Energy* **2012**, *91*, 366–374. [CrossRef]
16. Wang, F.; Fang, X.; Zhang, Z. Preparation of phase change material emulsions with good stability and little supercooling by using a mixed polymeric emulsifier for thermal energy storage. *Sol. Energy Mater. Sol. Cells* **2018**, *176*, 381–390. [CrossRef]
17. Wang, F.; Ling, Z.; Fang, X.; Zhang, Z. Optimization on the photo-thermal conversion performance of graphite nanoplatelets decorated phase change material emulsions. *Sol. Energy Mater. Sol. Cells* **2018**, *186*, 340–348. [CrossRef]
18. Wang, Y.; Chen, Z.; Ling, X. A molecular dynamics study of nano-encapsulated phase change material slurry. *Appl. Therm. Eng.* **2016**, *98*, 835–840. [CrossRef]
19. Kong, M.; Alvarado, J.L.; Terrell, W.; Thies, C. Performance characteristics of microencapsulated phase change material slurry in a helically coiled tube. *Int. J. Heat Mass Transf.* **2016**, *101*, 901–914. [CrossRef]
20. Dutkowski, K.; Fiuk, J.J. Experimental research of viscosity of microencapsulated PCM slurry at the phase change temperature. *Int. J. Heat Mass Transf.* **2019**, *134*, 1209–1217. [CrossRef]
21. Dutkowski, K.; Fiuk, J.J. Experimental investigation of the effects of mass fraction and temperature on the viscosity of microencapsulated PCM slurry. *Int. J. Heat Mass Transf.* **2018**, *126*, 390–399. [CrossRef]
22. Frank, A.J. TA Instruments Normal Stresses in Shear Flow. Available online: www.tainstruments.com/pdf/literature/AAN007e_Normal_forces_in_Shear_flow.pdf (accessed on 24 August 2019).
23. Macosko, C.W. *Rheology. Principles, Measurements and Applications*; Wiley-VCH: Weinheim, Germany, 1994; ISBN 1560815795.
24. Chatterjee, T.; Van Dyk, A.K.; Ginzburg, V.V.; Nakatani, A.I. Formulation-controlled positive and negative first normal stress differences in waterborne hydrophobically modified ethylene oxide urethane (HEUR)-latex suspensions. *ACS Macro Lett.* **2017**, *6*, 716–720. [CrossRef]
25. Aral, B.K.; Kalyon, D.M. Viscoelastic material functions of noncolloidal suspensions with spherical particles. *J. Rheol.* **1997**, *41*, 599–620. [CrossRef]
26. Zhang, T.; Lu, Q.; Lü, Y.; Wu, G. Determination of critical micelle concentration of sodium dodecyl sulfate in butyl acrylate emulsions. *Polym. Bull.* **2015**, *72*, 2215–2225. [CrossRef]
27. Helgeson, M.E. Colloidal behavior of nanoemulsions: Interactions, structure and rheology. *Curr. Opin. Colloid Interface Sci.* **2016**, *25*, 39–50. [CrossRef]
28. Gutiérrez, J.M.; Solè, I.; Maestro, A.; González, C.; Solans, C. Optimization of nano-emulsion preparation by low-energy methods in an ionic surfactant system. *Langmuir* **2006**, *22*, 8326–8332.
29. Tadros, T.; Izquierdo, P.; Esquena, J.; Solans, C. Formation and stability of nano-emulsions. *Adv. Colloid Interface Sci.* **2004**, *108–109*, 303–318. [CrossRef] [PubMed]
30. Estellé, P.; Cabaleiro, D.; Żyła, G.; Lugo, L.; Murshed, S.M.S. Current trends in surface tension and wetting behavior of nanofluids. *Renew. Sustain. Energy Rev.* **2018**, *94*, 931–944. [CrossRef]

31. Adzima, B.J.; Velankar, S.S. Pressure drops for droplet flows in microfluidic channels. *J. Micromech. Microeng* **2006**, *16*, 1504–1510. [CrossRef]
32. Cabaleiro, D.; Agresti, F.; Barison, S.; Marcos, M.A.; Prado, J.I.; Rossi, S.; Bobbo, S.; Fedele, L. Development of paraffinic phase change material nanoemulsions for thermal energy storage and transport in low-temperature applications. *Appl. Therm. Eng.* **2019**, *159*, 113868. [CrossRef]
33. Fedele, L.; Colla, L.; Bobbo, S.; Barison, S.; Agresti, F. Experimental stability analysis of different water- based nanofluids. *Nanoscale Res. Lett.* **2011**, *6*, 300. [CrossRef] [PubMed]
34. Halelfadl, S.; Estellé, P.; Aladag, B.; Doner, N.; Maré, T. Viscosity of carbon nanotubes water-based nanofluids: Influence of concentration and temperature. *Int. J. Therm. Sci.* **2013**, *71*, 111–117. [CrossRef]
35. Gómez-Villarejo, R.; Aguilar, T.; Hamze, S.; Estellé, P.; Navas, J. Experimental analysis of water-based nanofluids using boron nitride nanotubes with improved thermal properties. *J. Mol. Liq.* **2019**, *277*, 93–103. [CrossRef]
36. Hernaiz, M.; Alonso, V.; Estellé, P.; Wu, Z.; Sundén, B.; Doretti, L.; Mancin, S.; Çobanoğlu, N.; Karadeniz, Z.H.; Garmendia, N.; et al. The contact angle of nanofluids as thermophysical property. *J. Colloid Interface Sci.* **2019**, *547*, 393–406. [CrossRef] [PubMed]
37. COST Action NanoUptake Website. Available online: http://nanouptake.uji.es/ (accessed on 24 August 2019).
38. Vargaftik, N.B.; Volkov, B.N.; Voljak, L.D. International tables of surface tension of water with air. *J. Phy. Chem. Ref. Data* **1983**, *12*, 817–820. [CrossRef]
39. Żyła, G.; Fal, J.; Estellé, P. Thermophysical and dielectric profiles of ethylene glycol based titanium nitride (TiN–EG) nanofluids with various size of particles. *Int. J. Heat Mass Transf.* **2017**, *113*, 1189–1199. [CrossRef]
40. Lemmon, E.W.; Huber, M.L.; McLinden, M.O. NIST Standard Reference Database 23: Reference Fluid Thermodynamic and Transport Properties (REFPROP), Version 9.0. Physical and Chemical Properties; 2010. Available online: https://www.nist.gov/programs-projects/reference-fluid-thermodynamic-and-transport-properties-database-refprop (accessed on 24 August 2019).
41. Ferrer, G.; Gschwander, S.; Solé, A.; Barreneche, C.; Fernández, A.I.; Schossig, P.; Cabeza, L.F. Empirical equation to estimate viscosity of paraffin. *J. Energy Storage* **2017**, *11*, 154–161. [CrossRef]
42. Mikkola, V.; Puupponen, S.; Saari, K.; Ala-Nissila, T.; Seppälä, A. Thermal properties and convective heat transfer of phase changing paraffin nanofluids. *Int. J. Therm. Sci.* **2017**, *117*, 163–171. [CrossRef]
43. Murshed, S.M.S.; Estellé, P. A state of the art review on viscosity of nanofluids. *Renew. Sustain. Energy Rev.* **2017**, *76*, 1134–1152. [CrossRef]
44. Bullard, J.W.; Pauli, A.T.; Garboczi, E.J.; Martys, N.S. A comparison of viscosity–concentration relationships for emulsions. *J. Colloid Interface Sci.* **2009**, *330*, 186–193. [CrossRef] [PubMed]
45. Pal, R.; Rhodes, E. Viscosity/concentration relationships for emulsions viscosity. *J. Rheol.* **1989**, *33*, 1021–1045. [CrossRef]
46. Taylor, G.I. The viscosity of a fluid containing small drops of another fluid. *Philos. Trans. R. Soc. London Ser. A* **1932**, *138*, 41–48. [CrossRef]
47. Einstein, A. A new determination of molecular dimension. *Ann. Phys.* **1906**, *19*, 289–305. [CrossRef]
48. Bicerano, J.; Douglas, J.F.; Brune, D.A. Polymer model for the viscosity of particle dispersions model for the viscosity of particle dispersions. *J. Macromol. Sci. Part C* **1999**, *39*, 561–642. [CrossRef]
49. Pal, R. Modeling the viscosity of concentrated nanoemulsions and nanosuspensions. *Fluids* **2016**, *1*, 11. [CrossRef]
50. Pal, R. A new model for the viscosity of asphaltene solutions. *Can. J. Chem. Eng.* **2015**, *93*, 747–755. [CrossRef]
51. Pal, R. A new linear viscoelastic model for emulsions and suspensions. *Polym. Eng. Sci.* **2008**, *48*, 1250–1253. [CrossRef]
52. Lewis, T.B.; Nielsen, L.E.; Company, M. Dynamic mechanical properties of particulate-filled composites. *J. Appl. Polym. Sci.* **1970**, *14*, 1449–1471. [CrossRef]
53. Krieger, I.M.; Dougherty, T.J. A mechanism for non-Newtonian flow in suspensions of rigid spheres. *Trans. Soc. Rheol.* **1959**, *3*, 137–152. [CrossRef]
54. Brinkman, H.C. The viscosity of concentrated suspensions and solutions. *J. Chem. Phys.* **1952**, *20*, 571. [CrossRef]
55. Roscoe, R. The viscosity of suspensions of rigid spheres. *Br. J. Appl. Phys.* **1952**, *3*, 267–269. [CrossRef]
56. Chiesa, M.; Garg, J.; Kang, Y.T.; Chen, G. Thermal conductivity and viscosity of water-in-oil nanoemulsions. *Colloids Surfaces A Physicochem. Eng. Asp.* **2008**, *326*, 67–72. [CrossRef]

57. Xiang, N.; Yuan, Y.; Sun, L.; Cao, X.; Zhao, J. Simultaneous decrease in supercooling and enhancement of thermal conductivity of paraffin emulsion in medium temperature range with graphene as additive. *Thermochim. Acta* **2018**, *664*, 16–25. [CrossRef]
58. Safari, A.; Saidur, R.; Sulaiman, F.A.; Xu, Y.; Dong, J. A review on supercooling of phase change materials in thermal energy storage systems. *Renew. Sustain. Energy Rev.* **2017**, *70*, 905–919. [CrossRef]
59. Zeppieri, S.; Rodríguez, J.; López De Ramos, A.L. Interfacial tension of alkane + water systems. *J. Chem. Eng. Data* **2001**, *46*, 1086–1088. [CrossRef]
60. Goebel, A.; Lunkenheimer, K. Interfacial tension of the water/ n-alkane interface. *Langmuir* **2002**, *13*, 369–372. [CrossRef]
61. Oh, S.G.; Shah, D.O. Effect of counterions on the interfacial tension and emulsion droplet size in the oil/water/dodecyl sulfate system. *J. Phys. Chem.* **1993**, *97*, 284–286. [CrossRef]
62. Fainerman, V.B.; Lylyk, S.V.; Aksenenko, E.V.; Petkov, J.T.; Yorke, J.; Miller, R. Surface tension isotherms, adsorption dynamics and dilational visco-elasticity of sodium dodecyl sulphate solutions. *Colloids Surfaces A Physicochem. Eng. Asp.* **2010**, *354*, 8–15. [CrossRef]

Article

Wettability Control for Correct Thermophysical Properties Determination of Molten Salts and Their Nanofluids †

Yaroslav Grosu [1,*], Luis González-Fernández [1], Udayashankar Nithiyanantham [1,2] and Abdessamad Faik [1,*]

1 CIC Energigune, Albert Einstein, 4801510 Miñano, Spain; lgonzalez@cicenergigune.com (L.G.-F.); nudayashankar@cicenergigune.com (U.N.)

2 Applied Physics II Department, Faculty of Science and Technology, University of the Basque Country (UPV/EHU), PO Box 644, 48080 Bilbao, Spain

* Correspondence: ygrosu@cicenergigune.com (Y.G.); afaik@cicenergigune.com (A.F.); Tel.: +34-945-297-108 (Y.G.)

† This paper is an extended version of our paper published in 1st International Conference on Nanofluids (ICNf) and the 2nd European Symposium on Nanofluids (ESNf), Castello, Spain, 26–28 June 2019.

Received: 13 September 2019; Accepted: 30 September 2019; Published: 2 October 2019

Abstract: Proper recording of thermophysical properties for molten salts (MSs) and molten salts based nanofluids (MSBNs) is of paramount importance for the thermal energy storage (TES) technology at concentrated solar power (CSP) plants. However, it is recognized by scientific and industrial communities to be non-trivial, because of molten salts creeping (scaling) inside a measuring crucible or a sample container. Here two strategies are proposed to solve the creeping problem of MSs and MSBNs for the benefit of such techniques as differential scanning calorimetry (DSC) and laser flash apparatus (LFA). The first strategy is the use of crucibles with rough inner surface. It was found that only nanoscale roughness solves the creeping problem, while micron-scale roughness does not affect the wetting phenomena considerably. The second strategy is the use of crucible made of or coated with a low-surface energy material. Both strategies resulted in contact angle of molten salt higher than 90° and as a result, repeatable measurements in correspondence to the literature data. The proposed methods can be used for other characterization techniques where the creeping of molten salts brings the uncertainty or/and unrepeatability of the measurements.

Keywords: molten salt; thermophysical properties determination; wetting; differential scanning calorimetry; laser flash apparatus

---

1. Introduction

Molten salts are commonly used as a storage material at Concentrated Solar Power (CSP) plants [1,2], where their amount is of the order of thousands of tons [3–5]. For example, at GEMASOLAR CSP plant 8500 tons of Solar salt is used [3], at Andasol-1 CSP plant it is 28,500 tons [4] and at Solana CSP plant it is 125,000 tons [5]. Currently, so-called Solar salt (60–40 wt% of $NaNO_3$–KNO_3) is used most often. Hence, knowing the precise values of their heat capacity (Cp) and thermal conductivity (λ) is of the highest importance. However, it is recognized by the scientific and industrial communities to be very complicated, as reflected by the discrepancy of the published results. The same problem stands for molten salts based nanofluids [6]. In the last few years, the quality and accuracy of the thermophysical properties data of inorganic salts have been widely debated. This is due to the discrepancy of the published experimental results, obtained by different groups using different techniques [6]. As example, a round-robin test was performed in 2017 to measure Cp of the Solar salt by 11 institutions [7].

In order to get reliable values from such widely used techniques as Differential Scanning Calorimetry (DSC) or Laser Flash Apparatus (LFA), the sample must be in a good contact with the crucible. However, due to strong wetting typically molten salts creep on the walls of the measuring crucible Scheme 1. This results in a non-uniform thickness and an unacceptable error of the measurement.

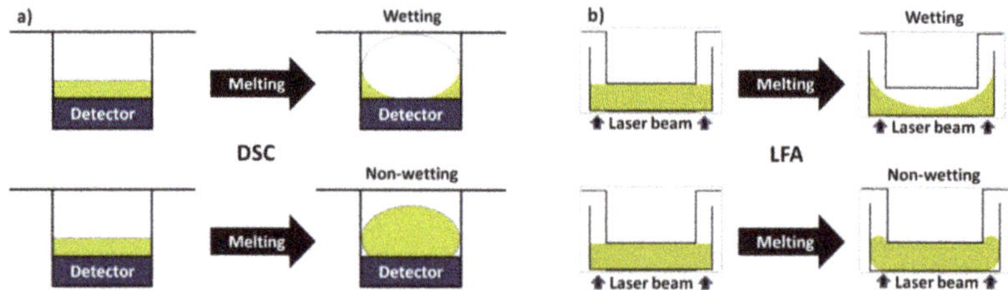

Scheme 1. Behavior of molten salt inside (**a**) DSC and (**b**) LFA crucible in case of wetting (top) and non-wetting (bottom) conditions.

The salt can be immobilized in the crucible if its quantity is large enough [8–10]. Such an approach was recently used by Muñoz-Sánchez et al. for Solar salt-based nanofluids [8]. Earlier, positive effect of larger mass in the DSC crucible on the accuracy of DSC measurements was noted by Thoms [9], and Hohne et al. [10]. However, the use of a larger mass has several important disadvantages. Firstly, due to thermal expansion of a larger quantity of molten salt, pressure builds up during the measurement in the measuring crucible. Obviously, in this case what is measured is not isobaric heat capacity (heat capacity at constant pressure). Besides, elevated pressure can lead to the leakage of the salt or even to the crucible deformation. This brings the risk of a critical damage of the equipment as a result of high corrosivity of molten salts at elevated temperature. Sometimes a hole is made in the crucible lead to prevent pressure increase. However, this can lead to the salt leakage from the overfilled crucible.

It should be noted that while wetting properties (like contact angle) of molten salts and molten salts nanofluids are widely discussed in the literature [11–13], their control to avoid creeping is rarely discussed. In particular, a round-robin test was reported in Reference [11] to identify the effect of nanoparticles doping on the contact angle of molten nitrate salt. The method of capillary rise was applied to molten carbonate salt and porous electrodes [12]. Wettability of carbon surfaces by molten chloride salts was studied in Reference [13].

In this work, we show that by using crucible made of a low-surface energy material or by decorating the interior crucible with nanoroughness increases the wetting contact angle and mitigates the wetting phenomena, which is responsible for displacement and creeping of the salt. We propose simple chemical or physical methods of introducing such roughness, which does not require special equipment and can be applied to already existing commercial measuring cells and crucibles. We demonstrated the strategy of low-surface energy material by testing several crucibles of standard design, but made of materials with low-surface energy. We believe that the proposed methodology can be extrapolated to other characterization techniques, where wetting phenomena result in molten salt displacement, such as laser flash, rheometry, etc., see Patent [14].

2. Materials and Methods

2.1. Materials

$NaNO_3$ and KNO_3 from SQM (Sociedad Química y Minera) were used in this work. These nitrates have a purity of ≥99%, with the list of impurities given elsewhere [12]. The Solar salt was prepared from pre-dried $NaNO_3$ and KNO_3 mixed in 60:40 wt% proportion. This mixture was melted at 360 °C and kept at this temperature during 6 h, followed by grinding in an agate mortar until homogeneous fine powder was obtained. Finally, 1 wt% of 50 nm Fe_2O_3 nanoparticles from Sigma-Aldrich were added to the mixture and subjected to the physical shaking method described elsewhere [15]. The Solar salt was prepared in a similar way, but without adding the nanoparticles. SPEX Sample Prep 9000-series High-Energy shaker was used for this purpose for the time period of 15 min using a quantity of 2 g of the mixture in a 50 mL aluminum bottle.

2.2. *Differential Scanning Calorimetry (DSC)*

Heat capacity (Cp) of the salts was measured by DSC technique (TA instruments Q2500) in a ramp mode with a heating rate of 10 °C/min. Sapphire and tin were used for the calibration materials. Around 10 mg samples were measured using aluminum crucibles, hermetically closed inside the glove-box under controlled argon atmosphere. Moreover, 3 types of T-zero hermetic Al-crucibles were used:

1. As received Al-DSC crucible from the supplier.
2. Partially leached Al-DSC crucible. These crucibles were prepared as by introducing 10 mL of 37% HCl acid into the pristine non-treated Al-DSC crucible for 15 min, followed by distilled water washing and drying at 60 °C.
3. Completely leached Al-DSC crucible was prepared the same way as the partially leached Al-DSC crucible, but by repeating the leaching procedure 5 times.

2.3. *Laser Flash Apparatus (LFA)*

Thermal diffusivity of the salts was measured by means of a Laser Flash Apparatus (Netzsch LFA 457 model). A graphite film prime over the top and bottom surfaces of the holder sample was applied aiming to increase the absorption and emission of radiation. Thermal conductivity of the nanofluids was calculated from thermal diffusivity (α), measured under non-equilibrium conditions, density (ρ) and specific heat capacity (Cp), according to the following Equation $\lambda(T) = \alpha(T)\cdot\rho(T)\cdot Cp(T)$. We note that in this formula the convective heat transfer is not taken into account, which in general can have an important contribution for nanofluids. This is, however, a reasonable neglection, as it is known that if Rayleigh number is less than 1400 [16] one can consider that the convection phenomenon is negligible and for the used configuration the Ra is far below 1 even with most conservative estimations. This is due to the very low thickness of the sample (below 1 mm), highly viscous sample, very small temperature difference between the top and the bottom of the crucible (below 1 °C) and very limited time of the LFA experiment.

Three types of LFA crucibles with identical design were used in this work:

1. As received PtRh-crucible from Netzsch.
2. Similar crucible, but custom-made from zinc (Zn-crucible).
3. Similar crucible, but custom-made from Stainless Steel (SS) 316-crucible. The interior of this crucible was also decorated with roughness by means of mechanical treatment.

2.4. *Scanning Electron Microscopy (SEM)*

Quanta 200 FEG SEM was used in vacuum mode with electron beam energies 10 kV, 20 kV and 30 kV with a Back Scattered Electron Detector (BSED) and an Everhart-Thornley Detector (ETD).

2.5. *Optical Microscope*

Carl Zeiss Axio optical microscope was used to scan the surface of the samples after the corrosion tests.

2.6. *Contact Angle*

The contact angle of molten salt on different surfaces was estimated by treating the optical images of 10 mg drop of molten salt at 300 °C. The images were obtained by Nikon D3300 digital camera.

The images were processed by ImageJ software in order to calculate the contact angle. At least 3 experiments were performed for each surface. At least 10 images were processed for each experiment. The highest standard deviation of the measurements was not exceeding ±7°. While such deviation is relatively high for modern equipment for contact angle measurements, it was sufficient to clearly capture the wetting-to-non-wetting transition.

2.7. Details on the Development of Crucibles with Roughness

In order to introduce roughness on the inner surface of the measuring crucibles electroetching, chemical etching and mechanical treatment methods were applied. The details on these methods are provided below. The crucibles were characterized by SEM and optical microscopy to identify the obtained roughness. Contact angle of molten salt on the surface of the crucibles was measured to verify whether the non-wetting condition was achieved.

2.7.1. Electroetching

Electroetching method was performed using an electric discharge machine ONA DATIC D360. By varying the applied voltage and the time of exposure different roughnesses were obtained (Figure 1a). Roughness in the range of 40–800 μm were achieved (Table 1). It was not possible to obtain nanoscale roughness by this method. Molten salt demonstrated a contact angle below 90° for all the obtained samples, therefore, the non-wetting condition was not reached.

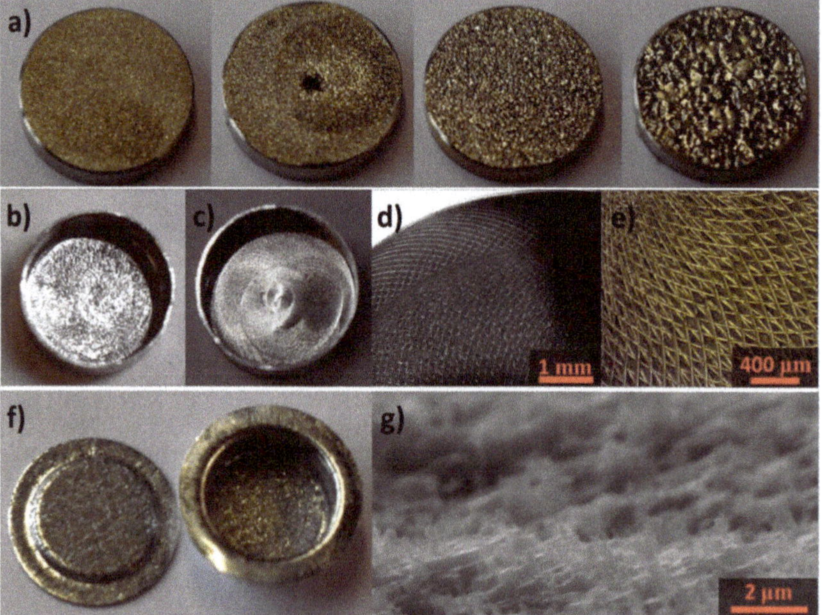

Figure 1. Optical and SEM images of different surfaces with introduced roughness. (**a**) SS316 with roughness obtained by electroleaching; (**b**) standard LFA PtRh-crucible with mechanically added roughness; (**c**) custom-made LFA SS316-crucible with mechanically added roughness and its (**d**) SEM and (**e**) optical images; (**f**) standard DSC T-zero hermetic Al-crucible with acid leaching roughness and its (**g**) SEM image.

Table 1. Roughness and non-wetting condition obtained by different methods.

Method	Scale of Roughness	Non-Wetting Condition
Electroetching 1	40 μm	No
Electroetching 2	110 μm	No
Electroetching 3	200 μm	No
Electroetching 4	800 μm	No
Mechanical treatment Dremel	800 nm + 10–20 μm	Yes
Mechanical treatment 2	700–1200 nm + 100 μm	Yes
Chemical etching	200–400 nm + 1–3 μm	Yes

2.7.2. Mechanical Treatment

Mechanical treatment of the crucibles was applied in order to obtain the surface roughness (Figure 1b–e). For this purpose, common commercial polishing tools were applied. In particular, surface treatment using Dremel 4000 tool with the Engraving cutter of 2.4 mm resulted in a bimodal roughness, namely, with scratches of around 800 nm wide with large recesses of around 10–20 μm (Table 1). Such roughness resulted in a non-wetting condition for the molten salt (see Section 3.2 for more details).

2.7.3. Chemical Etching

The chemical etching of the crucibles was achieved by exposing their inner surface to 37% HCl acid (Figure 1f). The optimization was carried out based on the exposure time, while providing fresh acid every 15 min. The crucible prepared by 15, 30, 45, 60, 75 min exposures were examined. It was found that independently on the time of etching the roughness on the scale of 200–400 nm was obtained (Figure 1g) with occasional large rectangle pores of 1–3 μm (Figure 2b and Table 1). On the other hand, the coverage of the surface with such roughness depends on the exposure time. It was identified that etching of 75 min resulted in a complete coverage of the surface with the roughness, while lower time resulted in partial coverage (Figure 2b). It was identified that such roughness results in the non-wetting condition of molten salts (Figure 2c).

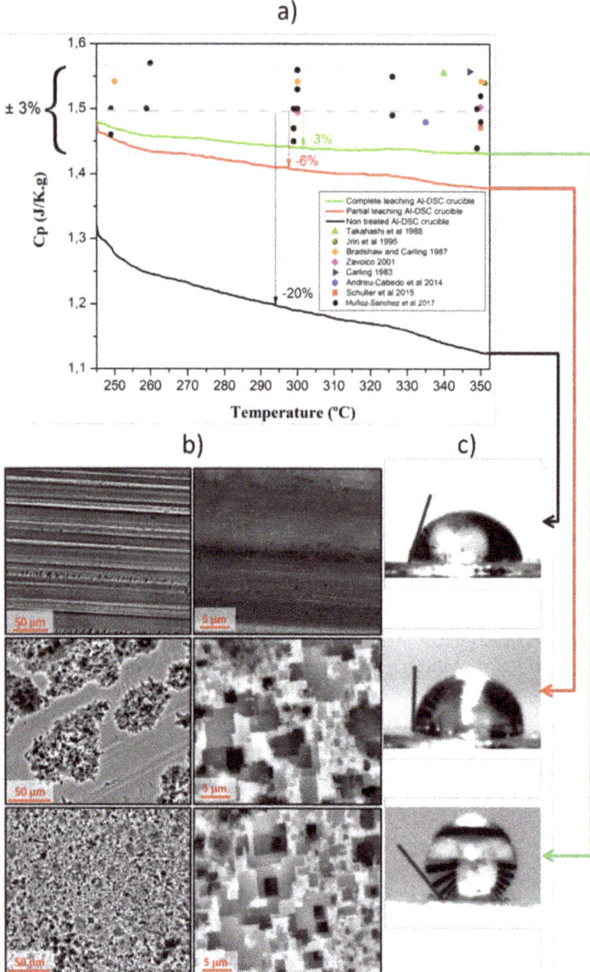

Figure 2. (**a**) Heat capacity of Solar salt, (**b**) SEM images of the crucible surface and (**c**) contact angle of Solar salt on the surface of the DSC crucible at 300 °C for the cases of pristine non-treated Al-DSC crucible (bottom), partially leached Al-DSC crucible (middle) and completely leached Al-DSC crucible (top).

3. Results

Scheme 1 demonstrates the problem of molten salt creeping inside a measuring crucible for such techniques as Differential Scanning Calorimetry (DSC) and Laser Flash Apparatus (LFA). As can be seen from Scheme 1, the wetting condition results in a molten salt creeping, eliminating the proper contact of the sample with the detector of the equipment. On the contrary, the non-wetting condition (contact angle of molten salt with the crucible is greater than 90°) would solve the creeping problem and results in a proper contact of the salt with the detector of the equipment. For most of the common materials the contact angle of molten salt is lower than 90° (Figure 3). Below we explore two strategies of wettability control of molten salts for such common techniques as DSC and LFA, namely the use of crucibles with microroughness and the use of crucibles made of (or coated with) low-surface energy materials. For this purpose, we explored different types of roughness (Figure 1), as well as used low surface energy materials (Figure 6) for custom-made crucibles (Figure 4).

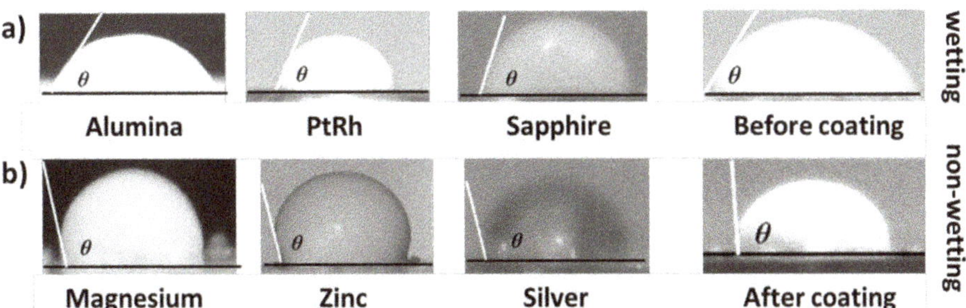

Figure 3. Contact angle of molten Solar salt at 300 °C on the surface of (**a**) alumina, PtRh-alloy, sapphire, SS316 (wetting condition) and on the surface of (**b**) magnesium, zinc, silver and SS316 coated with gold (non-wetting condition).

Figure 4. LFA crucible made of (**a**) PtRh-alloy (commercial) and (**b**) zinc (custom-made).

3.1. *Microroughness for Wettability Control of Molten Salts*

Surface roughness considerably affects the contact angle of a liquid resting on it [17–19]. In fact, introduction of microroughness is among one of the common methods to create super-lyophobic surfaces [17–19] surfaces, which are non-wetted by most of the liquids (contact angle more than 90°). Hence, introducing the surface roughness on the interior of the DSC or LFA crucibles can potentially mitigate the creeping problem.

In order to test such approach, different types of roughness were fabricated (Figure 1) and tested in terms of wetting properties with molten salts. First, we found that electroetching of stainless-steel materials (Figure 1a) results in a roughness, which does not bring the non-wetting condition for molten nitrate salt. Different roughness was obtained depending on the applied voltage and exposure time (Figure 1a), however, none of them resulted in the roughness capable of improving the wetting properties of molten nitrate salt. The main reason for such results was the absence of nanoroughness (while macroscopic roughness was obtained instead). On the contrary, nanoroughness was obtained by mechanical means by applying standard carving and polishing tools (Figure 1b,c). Optical and scanning electron microscopy (Figure 1d,e) confirmed the formation of microroughness, while the contact angle estimation (Figure 5b) clearly demonstrated the transition from wetting to non-wetting condition for the SS316 crucibles. Table 1 summarizes the relation between obtained roughness and wetting conditions.

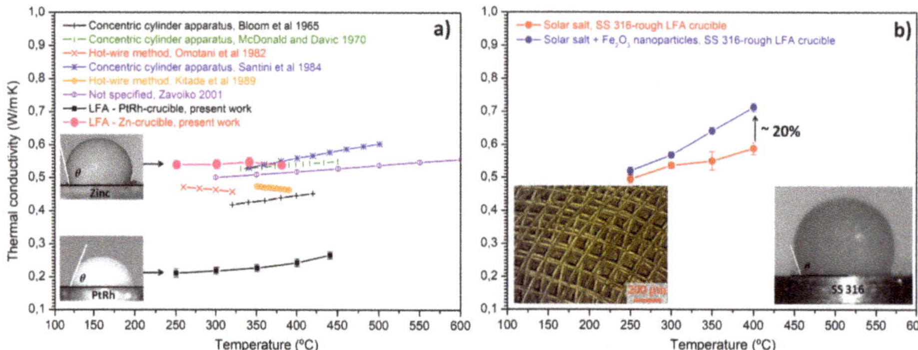

Figure 5. (**a**) Thermal conductivity of Solar salt measured by LFA using PtRh-and Zn-crucibles; Inserts: contact angles of Solar salt on the surface of the PtRh-and Zn-LFA crucibles. (**b**) Thermal conductivity of Solar salt and Solar salt-based nanofluid measured by LFA using rough SS 316-crucible; Inserts: contact angle of Solar salt on the surface of SS 316 crucible and optical image of the surface of SS 316 crucible.

For the standard T-zero hermetic Al-crucibles used for DSC analysis the acid leaching was found to be the most appropriate (Figure 1f). After the leaching, the inner surfaces of the DSC crucibles were decorated with a well-formed microroughness (Figure 1g).

Cp measurements for Solar salt are gathered in Figure 2. Al-crucibles with different roughness were used for these experiments. A clear correlation between surface roughness (Figure 2b), molten salt contact angel (Figure 2c) and Cp values (Figure 2a) was obtained. In particular, nanoroughness was obtained by acid leaching the surface of aluminum (Figure 2b). Such roughness strongly affects the contact angle of the molten salt (Figure 2c). It can be seen that, if the recommended by the supplier weight of the salt is used (10 mg), the use of the pristine non-treated Al-crucible deviate considerably from the values reported in the literature (Figure 2a). On the other hand, the deviation of the Cp values from the literature are no more than 3% when rough Al-crucible is used [7,20–26]. Partial roughness results in the values of Cp, which are between the cases of pristine and completely rough crucibles. One can also note, that when the pristine crucible is used, the Cp values decreases with temperature, apparently, due to the gradual creeping of the salt. In fact, Cp was found to decrease with each heating-cooling cycle. On the contrary, the slope of Cp is almost absent for the case of completely leached crucible (Figure 2a).

Similarly, it was found that the use of standard LFA PtRh-crucible for the recommended amount of Solar salt, results in values of thermal conductivity unacceptably lower as compared to the literature data (Figure 5a). On the other hand, the use of rough SS 316-crucible for LFA technique (Figure 1c–e) results in values of thermal conductivity for Solar salt in agreement with the literature (Figure 5b) and enhanced thermal conductivity for {Solar salt + 50 nm Fe_2O_3 nanoparticles} nanofluid as expected from the previous studies [6] Figure 5b. Such improvement correlated with the increase of the contact angle due to the introduced microroughness (Figure 5b).

It should be noted, that the observed wetting-to-non-wetting transition is not a trivial effect. In particular, a generally accepted Wenzel's model [27] predicts that roughness increases the observable contact angle θ_W only in case the Young´s contact angle on a smooth surface is greater than $\theta_Y > 90°$, while it reduces the observable contact angle θ_W, if $\theta_Y < 90°$ according to Equation (1).

$$cos\theta_W = r\, cos\theta_Y \tag{1}$$

where r is the roughness factor defined as the ratio of actual surface area over the projected surface area (basically the area without the roughness).

The contact angle of the molten nitrate salt is below 90° on the smooth surfaces of aluminum (Figure 2c, top), PtRd (Figure 5a, insert) and SS316. Therefore, according to the Wenzel´s model, roughness is expected to reduce the contact angle and enforce the creeping problem even more. This is in striking contrast to the observations, from which it is clearly seen that the contact angle was considerably increased due to the roughness (Figure 2c, bottom and Figure 5b, insert).

The explanation of the described controversy can be found within the Cassie–Baxter model [28]. This model predicts that non-wetting is enforced because air (or any other gas) is trapped inside the roughness (Equation (2)).

$$cos\theta_{CB} = r_f\, f\, cos\theta_Y + f - 1 \tag{2}$$

where θ_{CB} is the apparent contact angle according to the Cassie–Baxter model, r_f is the roughness factor of the wetted area, f is the area fraction of the projected wet area.

Generally, the Cassie–Baxter model is applied only to the cases when $\theta_Y > 90°$; however, it was hypothesized by Herminghaus that given the appropriate topology air-trapping can lead to the increased non-wetting (contact angle) even on surfaces with $\theta_Y < 90°$ [17]. This was later demonstrated experimentally for nanoporous gold surfaces [29] and for polycaprolactone nanopits [30]. We suggest that similar effect if observed in this work. Moreover, the preservation of the metastable state of entrapped gas inside the roughness is more probable in the case of molten salts as compared to the water, due to their much higher viscosity. As example, in our previous work it was demonstrated that the entrapped bubbles of air can remain in the molten salts based nanofluids for the periods of days depending on temperature [15,31], which is much longer than expected time of the DSC or LFA experiments.

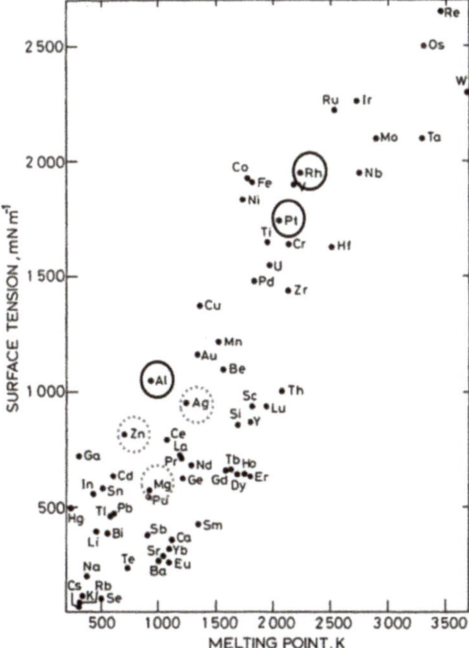

Figure 6. Surface tension of molten metals depending on their melting temperature [32].

Assuming that the hypothesis regarding the Cassie–Baxter state is true, one can use Equation (2) to estimate the effect of roughness on the contact angle and compare it with the experimental values. The parameters f and r_f can be estimated from the SEM characterization of the rough surfaces, assuming that there is no penetration of the salt into the pores. For the calculation we take θ_Y based on the contact angle measurements made on the pristine (smooth) surfaces of the crucibles. The results are summarized in Table 2. It can be seen that Equation (2) predicts the transition to the non-wetting condition ($\theta_{CB} > 90°$) in all the cases. In general, the model underestimates the values of the contact angle as compared to the experiment. We, however, note that due to the number of assumptions described above and relatively large error both for contact angle measurements as well as f and r_f determination, such comparison is done more for the qualitative prediction of the wetting-to-non-wetting transition, rather than for the qualitative comparison. Nevertheless, it gives a clear guidance for the optimization of the proposed method. Namely, the crucibles with smaller ratio between the surface exposed to the molten salt and the projected surface of the crucible, are expected to give more pronounced effect on the contact angle and, hence, are better for solving the creeping problem.

Table 2. Experimental and calculated values of the contact angles on the surfaces of different crucibles.

	Experiment Smooth	Experiment Rough	Calculated Rough	r_f	f
Al	71	128	118	4.8	0.21
PtRd	77	117	108	2.2	0.46
SS316	58	110	91	2.2	0.45

3.2. *Surface Energy for Wettability Control of Molten Salts*

The non-wetting condition for molten salt can also be reached by the use of low-surface energy materials for the measuring crucible. In particular, we found that for molten Solar salt at 300 °C the contact angle with gold surface is around 90° Figure 3b. This means that for all the materials with surface energy lower than that of gold, the contact angle of molten Solar salt at 300 °C is expected to be higher than 90°. This was demonstrated for the surfaces of magnesium, zinc and silver, Figure 3b. On the contrary, for all the materials with surface energy above the one of gold, the contact angle is expected to be below 90°. This was demonstrated for aluminum, PtRh-alloy, sapphire and stainless steel, Figure 3a. While surface energy may not be reported for all the materials due to complexity of such measurement, one can address more common property surface tension of metal at its melting point, which correlates with surface energy of this metal in a solid state. From Figure 6 [32] one can see that apart from magnesium, zinc and silver, there is a wide range of materials with surface tension lower than that of gold. In general, all those materials are expected to result in a contact angle of molten Solar salt higher than 90°. However, it is evident that metals with lower surface energy possess lower melting point, which limits their operational temperature range. Apart from lower melting temperature, one needs to take into account the possible corrosion issues.

Based on the analysis presented above a Zn-crucible for LFA was constructed using similar design as a standard PtRh-crucible provided by the supplier of the LFA equipment Figure 4. It can be seen that the use of Zn-crucible with low surface energy for LFA technique results in values of thermal conductivity, which is in good agreement with the literature (Figure 5a) [23,33–37]. On the contrary, much lower values are obtained when as received PtRh-crucible was used. The effect of the low surface energy of zinc on the contact angle of molten Solar salt is demonstrated in the inserts of Figure 5a, where the contact angle of molten Solar salt is below 90° for PtRh-crucible, while it is above 90° for Zn-crucible. It should be noted that the strategy of using low-surface energy materials for measuring crucibles can be easily adapted to a wide range of crucibles, as it does not requires a design modification, but it can be simply realized by coating an existing crucible with a low-surface energy material, like gold (see Figure 3, as an example).

In summary, it can be noted that the use of both strategies (nanoroughness and low-surface energy material) results in a non-wetting condition of the molten salt in the measuring crucibles. Such approach can be used for techniques other than DSC and LFA to deal with undesired wetting phenomenon. One can consider, as an example, the parallel plate rheometer, where according to the standard protocol all the gap between the parallel plates must be completely filled by the sample. If strong wetting takes place, a negative meniscus is formed, preventing the proper protocol execution. Therefore, reaching the non-wetting condition (positive meniscus) would have a positive effect for such technique [38].

4. Conclusions

Two strategies for controlling the wetting (creeping) of molten salts inside a measuring crucible were proposed in this work for the benefit of such widely used techniques as Differential Scanning Calorimetry (DSC) and Laser Flash Apparatus (LFA).

The first strategy suggests the use of crucibles with nanoroughness. Such roughness strongly increases the contact angle of a molten salt, preventing its creeping in the crucible. The proposed approach was tested using aluminum DSC crucibles by decorating their interior with nanoroughness achieved by acid leaching. LFA crucibles made of SS 316 with mechanically obtained roughness were also used. It was demonstrated that only nanoscale roughness results in a non-wetting condition for the molten salt, while micron-scale roughness did not considerably affect the contact angle of the molten salt. The main advantage of this method is that it can be chemically or physically applied to existing (commercial) crucibles, which ultimately results in a wide range of materials with proper compatibility and operational temperature range.

The second strategy implies the use of a crucible made of a low-surface energy material. This ultimately results in higher contact angle values. Zinc was used as a low-surface energy material to construct LFA crucibles and to prove the concept. The main advantage of this method is that no crucible design modifications are required. Moreover, it can be adapted to a wide range of measuring crucible by simply coating the existing crucible with a low-surface energy material.

The proposed strategies allowed recording the values of heat capacity and thermal conductivity of molten Solar salt in accordance with the literature. The proposed approach can be extended for the benefit of characterization techniques where displacement and creeping of molten salts bring an unacceptable error to the measurements.

5. Patents

There is one patent resulting from the work reported in this manuscript, namely, Reference [14].

Author Contributions: All authors contributed equally.

Funding: This research received no external funding.

Acknowledgments: This investigation is a contribution to the COST (European Cooperation in Science and Technology) Action CA15119: Overcoming Barriers to Nanofluids Market Uptake (NanoUptake). Y.G. acknowledges the EU COST Action CA15119: Overcoming Barriers to Nanofluids Market Uptake for financial support in the participation of the 1st International Conference on Nanofluids (ICNf) and the 2nd European Symposium on Nanofluids (ESNf) participation. The technical support from Yagmur Polat, Leticia Martinez and Cristina Luengo is highly appreciated.

Conflicts of Interest: The authors declare no conflicts of interest.

References

1. Vignarooban, K.; Xu, X.; Arvay, A.; Hsu, K.; Kannan, A.M. Heat transfer fluids for concentrating solar power systems—A review. *Appl. Energy* **2015**, *146*, 383–396. [CrossRef]
2. Gil, A.; Medrano, M.; Martorell, I.; Lázaro, A.; Dolado, P.; Zalba, B.; Cabeza, L.F. State of the art on high temperature thermal energy storage for power generation. Part 1—Concepts, materials and modellization. *Renew. Sustain. Energy Rev.* **2010**, *14*, 31–55. [CrossRef]

3. Burgaleta, J.I.; Arias, S.; Ramirez, D. Gemasolar, the first tower thermosolar commercial plant with molten salt storage. *Solarpaces* **2011**, *69*, 1–8.
4. Relloso, S.; Delgado, E.B. Experience with molten salt thermal storage in a commercial parabolic trough plant. In Andasol-1 commissioning and operation. In Proceedings of the 15th SolarPACES Symposium, Berlin, Germany, 14–18 September 2009.
5. Bonk, A.; Sau, S.; Uranga, N.; Hernaiz, M.; Bauer, T. Advanced heat transfer fluids for direct molten salt line-focusing CSP plants. *Prog. Energy Combust. Sci.* **2018**, *67*, 69–87. [CrossRef]
6. Arthur, O.; Karim, M.A. An investigation into the thermophysical and rheological properties of nanofluids for solar thermal applications. *Renew. Sustain. Energy Rev.* **2016**, *55*, 739–755. [CrossRef]
7. Muñoz-Sánchez, B.; Nieto-Maestre, J.; González-Aguilar, J.; Julia, J.E.; Navarrete, N.; Faik, A.; Bauer, T.; Bonk, A.; Navarro, M.E.; Ding, Y.; et al. Round robin test on the measurement of the specific heat of solar salt. *AIP Conf. Proc.* **2017**, *1850*, 080017.
8. Muñoz-Sánchez, B.; Nieto-Maestre, J.; Imbuluzqueta, G.; Marañón, I.; Iparraguirre-Torres, I.; García-Romero, A. A precise method to measure the specific heat of solar salt-based nanofluids. *J. Therm. Anal. Calorim.* **2017**, *129*, 905–914. [CrossRef]
9. Thoms, M.W. Adsorption at the Nanoparticle Interface for Increased Thermal Capacity in Solar Thermal Systems. Ph.D. Thesis, Massachusetts Institute of Technology, Cambridge, MA, USA, 28 June 2012.
10. Hohne, G.; Hemminger, W.F.; Flammersheim, H.-J. *Differential Scanning Calorimetry*, 2nd ed.; Springer-Verlag: Heidelberg, Germany, 2003; 298p, Available online: http://www.springer.com/la/book/9783540004677 (accessed on 3 September 2019).
11. Hernaiz, M.; Alonso, V.; Estellé, P.; Wu, Z.; Sundén, B.; Doretti, L.; Mancin, S.; Çobanoğlu, N.; Karadeniz, Z.H.; Garmendia, N.; et al. The contact angle of nanofluids as thermophysical property. *J. Colloid Interface Sci.* **2019**, *547*, 393–406. [CrossRef]
12. Lundblad, A.; Bergman, B. Determination of contact angle in porous molten-carbonate fuel-cell electrodes. *J. Electrochem. Soc.* **1997**, *144*, 984–987. [CrossRef]
13. Baumli, P.; Kaptay, G. Wettability of carbon surfaces by pure molten alkali chlorides and their penetration into a porous graphite substrate. *Mater. Sci. Eng. A* **2008**, *495*, 192–196. [CrossRef]
14. Faik, A.; Grosu, Y.; Udayashankar, N.; González-Fernández, L. Thermal Analysis Sample Container. European Patent application No 18382688.2-1003, 16 April 2019.
15. Grosu, Y.; Udayashankar, N.; Bondarchuk, O.; González-Fernández, L.; Faik, A. Unexpected effect of nanoparticles doping on the corrosivity of molten nitrate salt for thermal energy storage. *Sol. Energy Mater. Sol. Cells* **2018**, *178*, 91–97. [CrossRef]
16. Coquard, R.; Panel, B. Adaptation of the FLASH method to the measurement of the thermal conductivity of liquids or pasty materials. *Int. J. Therm. Sci.* **2009**, *48*, 747–760. [CrossRef]
17. Herminghaus, S. Roughness-induced non-wetting. *Europhys. Lett.* **2000**, *52*, 165–170. [CrossRef]
18. Quéré, D. Wetting and roughness. *Annu. Rev. Mater. Res.* **2008**, *38*, 71–99. [CrossRef]
19. Liu, T.; Kim, C.J. Turning a surface superrepellent even to completely wetting liquids. *Science* **2014**, *346*, 1096–1100. [CrossRef] [PubMed]
20. Takahashi, Y.; Sakamoto, R.; Kamimoto, M. Heat capacities and latent heats of $LiNO_3$, $NaNO_3$, and KNO_3. *Int. J. Thermophys.* **1988**, *9*, 1081–1090. [CrossRef]
21. Jriri, T.; Rogez, J.; Bergman, C.; Mathieu, J.C. Thermodynamic study of the condensed phases of $NaNO_3$, KNO_3 and $CsNO_3$ and their transitions. *Thermochim. Acta* **1995**, *266*, 147–161. [CrossRef]
22. Bradshaw, R.W.; Carling, R.W. *A Review of the Chemical and Physical Properties of Molten Alkali Nitrate Salts and Their Effect on Materials Used for Solar Central Receivers*; Report SAND87-8005; Sandia National Laboratories: Albuquerque, NM, USA, 1987.
23. Zavoico, A.B. *Design Basis Document*; SAND2001-2100; Sandia National Laboratories: Albuquerque, NM, USA, 2001.
24. Carling, R.W. Heat capacities of $NaNO_3$ and KNO_3 from 350 to 800 K. *Thermochim. Acta* **1983**, *60*, 265–275. [CrossRef]
25. Andreu-Cabelo, P.; Mondragon, R.; Hernandez, L.; Martines-Cuenca, R.; Cabedo, L.; Julia, J.E. Increment of specific heat of Solar Salt with SiO_2 nanoparticles. *Nanoscale Res. Lett.* **2014**, *9*, 1–9.
26. Schuller, M.; Shao, Q.; Lalk, T. Experimental investigation of the specific heat of a nitrate-alumina nanofluid for solar thermal energy storage systems. *Int. J. Therm. Sci.* **2015**, *91*, 142–145. [CrossRef]

27. Wenzel, R.N. Resistance of solid surfaces to wetting by water. *Ind. Eng. Chem.* **1936**, *28*, 988–994. [CrossRef]
28. Cassie, A.B.D.; Baxter, S. Wettability of porous surfaces. *Trans. Faraday Soc.* **1944**, *40*, 546–551. [CrossRef]
29. Abdelsalam, M.E.; Bartlett, P.N.; Kelf, T.; Baumberg, J. Wetting of regularly structured gold surfaces. *Langmuir* **2005**, *21*, 1753–1757. [CrossRef] [PubMed]
30. Martines, E.; Seunarine, K.; Morgan, H.; Gadegaard, N.; Wilkinson, C.D.; Riehle, M.O. Air-trapping on biocompatible nanopatterns. *Langmuir* **2006**, *22*, 11230–11233. [CrossRef] [PubMed]
31. Nithiyanantham, U.; Grosu, Y.; González-Fernández, L.; Zaki, A.; Igartua, J.M.; Faik, A. Corrosion aspects of molten nitrate salt-based nanofluids for thermal energy storage applications. *Sol. Energy* **2019**, *189*, 219–227. [CrossRef]
32. Keene, B.J. Review of data for the surface tension of pure metals. *Int. Mater. Rev.* **1993**, *38*, 157–192. [CrossRef]
33. Tricklebank, S.B. Molten Salt Mixtures. IX. The Thermal Conductivities of Molten Nitrate Systems. *Aust. J. Chem.* **1965**, *18*, 1171–1176.
34. McDonald, J.; Davis, H.T. Thermal conductivity of binary mixtures of alkali nitrates. *J. Phys. Chem.* **1970**, *74*, 725–730. [CrossRef]
35. Omotani, T.; Nagasaka, Y.; Nagashima, A. Measurement of the thermal conductivity of KNO_3-$NaNO_3$ mixtures using a transient hot-wire method with a liquid metal in a capillary probe. *Int. J. Thermophys.* **1982**, *3*, 17–26. [CrossRef]
36. Santini, R.; Tadrist, L.; Pantaloni, J.; Cerisier, P. Measurement of thermal conductivity of molten salts in the range 100–500 °C. *Int. J. Heat Mass Transf.* **1984**, *27*, 623–626. [CrossRef]
37. Kitade, S.; Kobayashi, Y.; Nagasaka, Y.; Nagashima, A. Measurement of the thermal conductivity of molten KNO_3 and $NaNO_3$ by the transient hot-wire method with ceramic-coated probes. *High Temp. High Press.* **1989**, *21*, 219–224.
38. Grosu, Y.; González-Fernández, L.; Nithiyanantham, U.; Faik, A. Wettability control for correct thermophysical properties determination of molten salts and their nanofluids. In Proceedings of the 1st International Conference on Nanofluids (ICNf), Castello, Spain, 26–28 June 2019.

Article

Magnetic Field Effect on Thermal, Dielectric, and Viscous Properties of a Transformer Oil-Based Magnetic Nanofluid †

Michal Rajnak [1,2,*], Zan Wu [3], Bystrik Dolnik [2], Katarina Paulovicova [1], Jana Tothova [2], Roman Cimbala [2], Juraj Kurimský [2], Peter Kopcansky [1], Bengt Sunden [3], Lars Wadsö [4] and Milan Timko [1,*]

1 Institute of Experimental Physics SAS, Watsonova 47, 04001 Kosice, Slovakia; paulovic@saske.sk (K.P.); kopcan@saske.sk (P.K.)
2 Faculty of Electrical Engineering and Informatics, Technical University of Kosice, Letna 9, 04200 Kosice, Slovakia; bystrik.dolnik@tuke.sk (B.D.); jana.tothova@tuke.sk (J.T.); roman.cimbala@tuke.sk (R.C.); juraj.kurimsky@tuke.sk (J.K.)
3 Department of Energy Sciences, Lund University, 22100 Lund, Sweden; Zan.Wu@energy.lth.se (Z.W.); Bengt.Sunden@energy.lth.se (B.S.)
4 Division of Building Materials, Lund University, 22100 Lund, Sweden; lars.wadso@byggtek.lth.se
* Correspondence: rajnak@saske.sk (M.R.); timko@saske.sk (M.T.)
† This paper is an extended version of our paper published in Conference proceedings of 1st International Conference on Nanofluids (ICNf2019), ISBN eBook en PDF: 978-84-685-3917-1, pp. 95–98.

Received: 30 October 2019; Accepted: 25 November 2019; Published: 28 November 2019

Abstract: Progress in electrical engineering puts a greater demand on the cooling and insulating properties of liquid media, such as transformer oils. To enhance their performance, researchers develop various nanofluids based on transformer oils. In this study, we focus on novel commercial transformer oil and a magnetic nanofluid containing iron oxide nanoparticles. Three key properties are experimentally investigated in this paper. Thermal conductivity was studied by a transient plane source method dependent on the magnetic volume fraction and external magnetic field. It is shown that the classical effective medium theory, such as the Maxwell model, fails to explain the obtained results. We highlight the importance of the magnetic field distribution and the location of the thermal conductivity sensor in the analysis of the anisotropic thermal conductivity. Dielectric permittivity of the magnetic nanofluid, dependent on electric field frequency and magnetic volume fraction, was measured by an LCR meter. The measurements were carried out in thin sample cells yielding unusual magneto-dielectric anisotropy, which was dependent on the magnetic volume fraction. Finally, the viscosity of the studied magnetic fluid was experimentally studied by means of a rheometer with a magneto-rheological device. The measurements proved the magneto-viscous effect, which intensifies with increasing magnetic volume fraction.

Keywords: magnetic nanofluid; magnetic nanoparticles; thermal conductivity; viscosity; permittivity

---

1. Introduction

Electrical equipment, like power transformers, inherently operate with energy losses, which lead to rises in temperature. To prevent premature aging and the ultimate failure of the equipment, the heat generated through energy loss must be dissipated. In practice, this is usually achieved by circulating transformer oils (TO), which also ensures the electrical insulation of energized conductors [1]. However, with the development of the future high-voltage network and smart grid, the increasing requirement for reliability and performance of dielectric and cooling liquids motivates researchers to develop more effective liquid media using nanotechnology [2]. Thus, various nanofluids have been synthesized with the aim of improving the cooling and insulating properties of the base dielectric liquid. Besides the conventional nanoparticles, such as TiO_2 or Al_2O_3, these improvements have been achieved also with graphene nanosheets [3,4], fullerene [5], or magnetic nanoparticles (MNPs) [6,7]. The stable dispersions of MNPs in TO (well-known as magnetic nanofluids or ferrofluids) have an advantage over other nanofluids due to the controllability of thermal, dielectric, and viscous properties by an external magnetic field. Like in other nanofluids, the addition of MNPs to TO causes an increase in viscosity and this is a drawback for pumping power. However, the phenomenon of thermomagnetic convection [8–10] and the possibility of controlling nanofluid flow by the external magnetic field makes magnetic nanofluids (MNFs) attractive and suitable candidates for the replacement of TO.

With regards to the potential application of transformer oil-based MNFs in power transformers, most research studies have focused on three key properties: thermal, dielectric, and viscous. Within the study of thermal properties, the thermal conductivity of MNFs attracts scientific interest, in addition to research on thermomagnetic convection. In comparison to TO, the thermal conductivity of MNFs is increased due to the presence of MNPs. This is often interpreted by means of various static and dynamic mechanisms and models used for nanofluids in general [11]. On the other hand, this property is a function of the base liquid, dispersed particles, size distribution, volume fraction, surfactant, and the external magnetic field. The influence of listed factors on the thermal conductivity of MNFs has been documented in numerous papers, some of them reviewed in a recent review article [12]. From the literature, one can see that the thermal conductivity of MNFs is generally enhanced in the presence of a magnetic field regardless of the MNF type. However, some of the studies report huge increments in thermal conductivity [13–15], while others refer to smaller increments, with the applied magnetic field [16,17]. These differences are also found for similar types of MNF samples and similar measurement setups used. One can assume that these discrepancies may originate in MNF characterization or in the differences in magnetic field distribution acting on studied MNFs. Generally, it was shown that such discrepancies might also arise due to nanoparticle aggregation [18]. It is well-known that under an external magnetic field, the nanoparticles in an MNF can exhibit self-assembly and micromotion [19]. Structural changes are initiated by moments of force causing the rotation of the nanoparticles that are parallel with the field. The interparticle magnetic forces can then lead to the formation of chain-like clusters [20]. Due to the formed heterogeneous structure in MNF, the thermal conductivity is anisotropic. It was found that the thermal conductivities parallel to the magnetic field direction are, in most cases, much higher than the perpendicular ones [21–23]. Thus, the magnetic field direction and distribution in the study of MNF thermal conductivity plays a key role.

The insulating and dielectric properties of TO were found to exhibit peculiar enhancement upon the addition of MNPs [23–27]. A comprehensive mechanism of the increased breakdown field strength is still unknown, however, a few models have been proposed. In the nanoparticle charging model [28], the difference in the dielectric permittivity of the dispersed nanoparticles and the surrounding oil is considered as an essential condition leading to polarization, charge trapping, and subsequent reduction of streamer velocity. Thus, the dielectric permittivity of MNF is a crucial parameter that has been extensively studied [29–31]. Again, the effect of an external magnetic field and the induced heterogeneous nanoparticle structure results in magneto-dielectric anisotropy [32]. This is mostly characterized by higher dielectric permittivity measured in the parallel configuration of electric and magnetic fields as compared to perpendicular configuration. Furthermore, the magnetic field induced MNP assembly also results in the well-known magneto-viscous effect. This effect is expressed as the ratio between the viscosity increment under the field and the viscosity value in the absence of the field and it reaches a maximum for a critical MNP size [33]. There are numerous studies demonstrating that the viscosity of MNFs is dependent on the flow shear rate, nanoparticle volume fraction, temperature, magnetic field, or MNP interaction [34–36]. Recently, an empirical correlation was proposed to predict magneto-viscous behavior of MNFs, as a function of these parameters [37].

In this study, we investigate the three key quantities of an MNF based on a novel, commercially available TO. Taking into account the referenced discrepancies in the thermal conductivity data in the literature, we intend to present other seemingly counter-intuitive experimental results. We discuss that the discrepancies may result from the way the magnetic field is applied and distributed through the MNF sample volume with regards to the measuring sensor. Similarly, we report on the unusual magneto-dielectric anisotropy and complement the study with the measurements of the viscosity of the MNF dependent on the magnetic field and magnetic volume fraction.

2. Materials and Methods

The investigated MNF is based on a commercially available inhibited insulating transformer oil MOL TO 40A. Its basic physical properties provided by the manufacturer are as follows: density 0.867 g/cm^3, kinematic viscosity 22 mm^2/s, pour point 228 K, flash point 413 K, and interfacial tension 42 mN/m. The dispersive phase consists of superparamagnetic iron oxide nanoparticles, which are coprecipitated from ferrous and ferric ions in an aqueous solution according to this well-known procedure [38]. To stabilize the MNP in the TO, we used oleic acid as a surfactant. The stabilized MNPs are dispersed in the TO in the following solid volume fraction 0.39%, 0.97%, 1.39%, 1.92%, and 3.62% with the corresponding physical parameters presented in Table 1. These samples have been stable for 2 years.

Table 1. The basic physical parameters of the prepared magnetic nanofluid samples.

Sample	Magnetization of Saturation (kA/m)	Magnetic Volume Fraction (%)	DC Magnetic Susceptibility (-)	Density (g/cm^3)	Solid Volume Fraction (%)
MOL TO 40A	-	-	-	0.867	0
MNF_MOL 1	1.68	0.377	0.05	0.884	0.39
MNF_MOL 2	3.598	0.807	0.11	0.909	0.97
MNF_MOL 3	5.223	1.171	0.15	0.927	1.39
MNF_MOL 4	7.469	1.675	0.21	0.95	1.92
MNF_MOL 5	9.802	2.198	0.24	1.023	3.62

The solid volume fraction is calculated according to the following formula: $\varphi = (\rho_{MNF} - \rho_{TO})/(\rho_{MNP} - \rho_{TO})$, where ρ_{MNF}, ρ_{TO}, and ρ_{MNP} are the densities of the MNF, TO, and MNP, respectively. The density of magnetite (5.18 g/cm^3) was substituted in ρ_{MNP}. The magnetic volume fraction is determined as a ratio of the MNF magnetization of saturation to the domain magnetization of magnetite (446 kA/m) [39]. The magnetic properties of the MNF samples were measured by means of a vibrating sample magnetometer installed on a cryogen-free superconducting magnet from Cryogenic Limited. The obtained magnetization curves measured at 298 K in the field ranging up to 6 T are shown in Figure 1. The mean particle size derived from the fitting of the magnetization curve by the superposition of Langevin functions is 11.1 nm. From Figure 1, one can observe the typical superparamagnetic behavior with zero coercivity and remanence. At higher magnetic fields (above 2 T), the magnetization is well saturated and increases with the increasing magnetic volume fraction.

The measurements of thermal conductivity were carried out by a thermal constant analyzer (TPS 2500S, Hot Disk AB, Sweden) using a transient plane source method (TPS). The spiral sensor with the axial and radial probing depth of 1.2 mm was inserted vertically into a Teflon container filled with the probed MNF. The diameter of the sensor is 3.2 mm. Figure 2 shows a schematic illustration and a real picture of the double Hot Disk sensor and its assembly with the Teflon sample container.

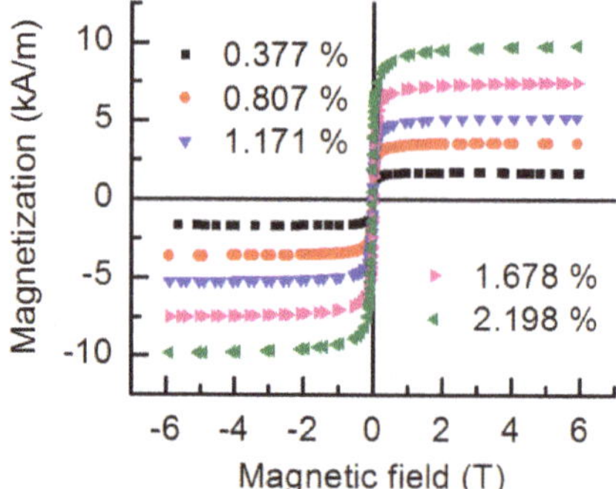

Figure 1. Magnetization curves of magnetic nanofluid samples with various magnetic volume fractions. The measurements were performed at a temperature of 298 K.

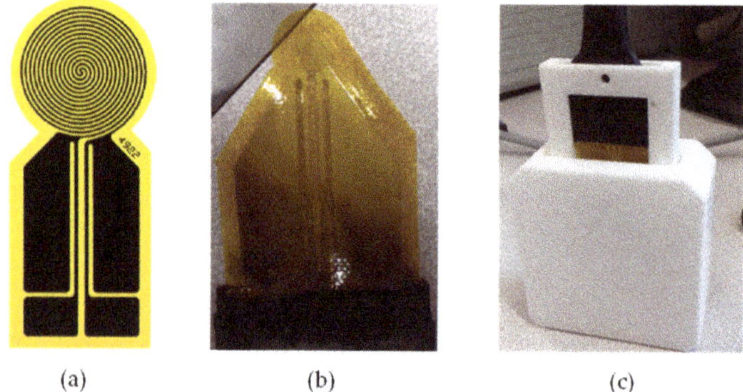

Figure 2. A schematic illustration of the double spiral Hot Disk sensor (**a**). The picture of the sensor with Kapton insulation (**b**). The sensor fixed in a Teflon frame inserted in the sample container (**c**).

The static magnetic field acting on the investigated MNF samples was generated between a pair of permanent magnets (37 mm apart) attached to the walls of the sample container (Figure 3a). With various sets of permanent magnets, the average magnetic field of 45 mT, 90 mT, and 210 mT in the sensor location was proven by a Hall probe. The particular magnetic field value was changed by manual attachment of the prepared magnetic set. Further experimental tests confirmed the unmeasurable effect of the present magnetic field on the Hot Disk sensor. In Figure 3b, we present a schematic arrangement of the permanent magnets with the simulated distribution of the magnetic flux density. A more detailed color pattern of magnetic flux density simulated in the sample area between the two strongest magnets is demonstrated in Figure 3c,d.

The numerical simulation of the magnetic flux density near the permanent magnets was simulated by Finite Element Method Magnetics (FEMM 4.2) using the triangular element mesh and the iterative conjugate solver. We changed the default mesh density to a finer mesh density, 1 mm for the air, and 0.25 mm for the permanent magnet. The electrical and magnetic parameters of the Neodymium Iron Boron permanent magnet was set as follows: the relative permeability $\mu_r = 1.049$, the coercivity $H_c = 1.13$ MA/m, and the electrical conductivity $\sigma = 0.667$ MS/m. The distribution of magnetic flux density and the isolines of magnetic flux density between the two permanent magnets was achieved by using the GNU Octave high-level language, version 4.0.0. Using this type of magnetic field distribution simulation supports subsequent analysis of the magnetic field dependent thermal conductivity of magnetic fluids.

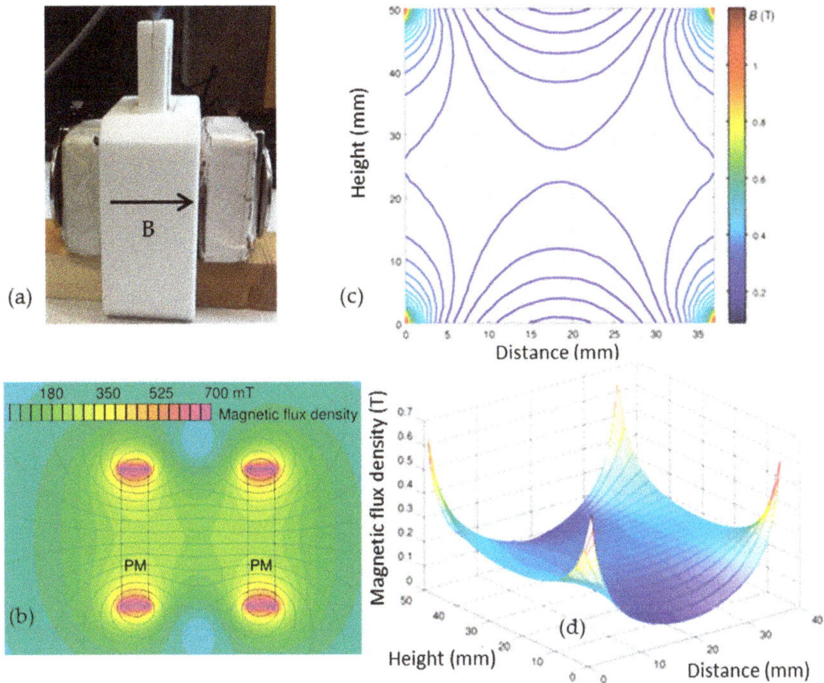

Figure 3. Set of permanent magnets attached to the Teflon sample container (**a**). Simulation of magnetic flux density distribution around the permanent magnets (PM) (**b**). More detailed color flux density plots for the sample area between the two strongest magnets applied in the experiment (**c**,**d**).

The frequency-dependent dielectric permittivity of the studied MNF was measured in commercial liquid crystal cells. The sandwiched type cells are composed of two flat glass pieces coated with indium tin oxide (ITO) conductive layers acting as electrodes. The electrodes are 50 μm apart while the active electrode area is 25 (5 × 5) mm^2. Firstly, the capacity of the empty cells C_0 was measured, yielding the value 7.5 pF. Then, the cells were filled with investigated TO and MNF samples. To study the magnetic field effect, the sample cell was placed between two poles of an electromagnet. The dielectric spectroscopy measurements were carried out at room temperature using an LCR meter (Agilent E4980A) with the effective voltage value across the capacitor of 1 V.

The viscous properties of the MNF samples under a magnetic field were studied by using an Anton Paar rheometer MCR 502 with the magneto-rheological device (MRD). MRD is a special accessory of the MCR rheometer, which enables all rheological tests with the simultaneous action of the external magnetic field.

3. Results and Discussion

3.1. Thermal Conductivity

In Figure 4a, we present the dependence of the thermal conductivity (TC) of the MNF on the magnetic volume fraction in the absence of the external magnetic field. The TC of the pure TO was found to be 0.127 W/mK. It is clearly shown that with the addition of MNP, the TC increases moderately. The lowest magnetic volume fraction 0.377% results in a 3.9% increase in TC, while for the highest magnetic volume fraction of 2.198%, the TC of MNF increases by 9.2%, compared to the pure TO. Clearly, the thermal conductivity increases with the magnetic volume fraction because the particle thermal conductivity is higher than that of the oil. Higher thermal conductivity of the particles indicates a lower temperature drop across the particles. In this way, various approaches could be chosen to model the experimentally well-known increase in TC with increasing MNP volume fraction. In the simple approach, taking into account the non-interacting spherical MNPs of low concentration, one can apply an effective medium theory, such as the Maxwell model described by Equation [11],

$$k_{\text{eff}} = \frac{k_{\text{p}} + 2k_{\text{f}} + 2\Phi\left(k_{\text{p}} - k_{\text{f}}\right)}{k_{\text{p}} + 2k_{\text{f}} - \Phi\left(k_{\text{p}} - k_{\text{f}}\right)} k_{\text{f}}, \tag{1}$$

where k and Φ denote the thermal conductivity and the solid volume fraction of the MNP. The subscripts eff, p, and f indicate the nanofluid, the particle, and the base fluid (TO), respectively. The substitution of particular values (for magnetite nanoparticles k_{p}= 1.39 W/mK is considered) yields k_{eff} of 0.29 W/mK and 1.24 W/mK for magnetic volume fractions equal to 0.377% and 0.807%, respectively. These values are higher than those measured in the experiments. Moreover, for higher MNP volume fractions, the relation yields negative values of k_{eff}. Thus, the Maxwell model is not sufficient to explain the observed thermal conductivity in the MNF. More sophisticated models with an emphasis on the interfacial thermal resistance (Kapitza resistance), the nanoscale layer between the TO and MNPs, or the nanoconvection induced by the Brownian motion of MNPs should be considered in a future study. The development of such a model is beyond the scope of this work.

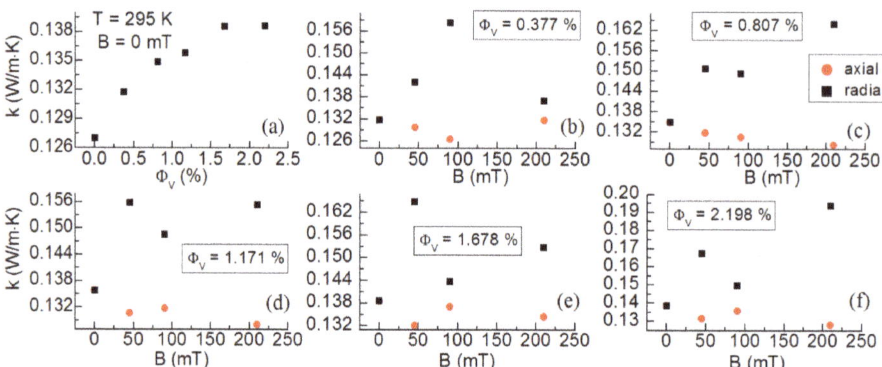

Figure 4. Dependence of the magnetic nanofluid (MNF) thermal conductivity on the magnetic volume fraction (**a**). The effect of the magnetic field on the thermal conductivity of the MNF of various magnetic volume fractions probed in the axial and radial directions of the Hot Disk sensor (**b–f**).

From Figure 4b–f, one can observe the changes of TC with an increasing magnetic field for particular MNF samples. These measurements were conducted in an anisotropic mode. First of all, when comparing the two components of TC, each MNF sample exhibits a higher TC in the radial direction (perpendicular to the magnetic field intensity), compared to the axial one (parallel with the magnetic field intensity). This is observed for each magnetic field value. The finding is partly in contradiction to the previous studies, which mostly report an opposite anisotropic effect. Usually, a higher TC in a parallel direction to the applied magnetic field is associated with MNP interaction with the field and the subsequent aggregation of MNP. However, one has to bear in mind that in these experiments, not only the magnetic field direction and intensity affect the aggregation and the TC of MNF, but the magnetic field gradient ∇B has a determining character too. In a homogenous magnetic field with $\nabla B = 0$, the nanoparticles do not experience any drag force. Nevertheless, the magnetic moments orientate (Brownian or Néel mechanism) with the field, and a chain-like structure may be formed. Such MNP structures locally form rapid heat conduction paths (percolation paths), and so along the paths, the TC is increased. The enhanced TC due to the formed aggregation has been proven numerically for nanofluids in general [40,41]. However, in a strong gradient magnetic field the formed MNP structures can weaken, break, and drag in the field gradient. It is very important to know the exact magnetic field distribution with regard to the sensor location in the MNF. From our simulations of the magnetic flux density in Figure 3, we can clearly observe denser contours of the magnetic field near the edges of the magnets. The increasing magnetic flux density towards the container edges are also illustrated in the color scale. The simulation, therefore, proves that the MNF in the experimental container experiences a gradient magnetic field, which exerts a drag force on the MNPs. As a result, instead of horizontal chain structures, the nanoparticles are forced to form aggregates directed to the edges of the container. Now, realizing that the thin Hot Disk sensor is located in the center of the container, it is clear that in the axial (horizontal) probing depth, the MNF contains fewer MNPs (decreased effective magnetic volume fraction) due to the above-mentioned reasoning. This is reflected in lower axial TC detected for each magnetic field value. On the other hand, the aggregates formed from the center towards the container edges constitute the increased heat conduction paths and contribute to the higher radial TC, as revealed in the experiment. However, in Figure 4b–f, we do not observe a strictly proportional increase of the radial TC with an increasing magnetic field. The data exhibits a rather stochastic behavior. In our opinion, this is attributed to a different re-distribution of MNPs at a new magnetic field value. As mentioned in Section 2, the new magnetic field value is achieved by attaching a new set of permanent magnets manually. Clearly, the manual attachment of the opposing magnets might not be exactly symmetric with regards to the Hot Disk sensor, and each set of magnets may lead to slightly different distributions of the MNP aggregates. Nevertheless, for each magnet set and the related magnetic field value, we have proven a higher radial TC in comparison to its axial part.

3.2. Dielectric Permittivity

From a dielectric point of view, we focus on the dielectric response of the studied MNF and present the dielectric permittivity measured from 700 Hz up to 100 kHz. This frequency range is chosen deliberately to observe magneto-dielectric anisotropy in MNFs that are free of pronounced interfacial polarization, conductivity and space charge migration, and electrode polarization, which has been documented, for example, in the paper with the reference number [42]. The resulting dielectric spectra for the three MNF samples are plotted in Figure 5.

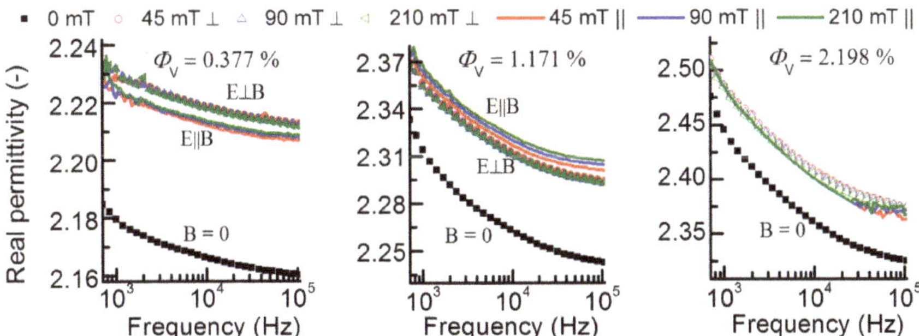

Figure 5. The spectra of real dielectric permittivity of magnetic nanofluids with various magnetic volume fractions. The graphs show the effect of the magnetic fields and their orientation with regards to measuring the electric field. The measurements were performed at room temperature.

The dielectric permittivity of the pure TO at 1 kHz is 2.12, and this value has been found to be quasi constant in the whole frequency range. The addition of MNP with the volume fraction of 0.377%, 1.171%, and 2.198% results in the permittivity values of 2.18, 2.31, and 2.45, respectively. Thus, at 1 kHz, these relatively low MNP volume fractions do not increase the oil's permittivity to a great extent. However, the spectrum of the permittivity is markedly shifted to higher values upon the application of the magnetic field, as demonstrated in the graphs of Figure 5. This effect is associated with MNP assembly in the field direction. For the sample of a higher magnetic volume fraction (2.198%), we do not observe any magneto-dielectric anisotropy, as the permittivity values measured in the parallel configuration of electric and magnetic field intensities coincide with those measured in the perpendicular configuration. This behavior is ascribed to measuring capacitor geometry (50 μm electrode separation distance, 25 mm^2 electrode area) and close packing of the MNP aggregates. In contrast to the gradient magnetic field in the previous case, in this experiment the magnetic field in the active sample volume is homogenous. It is believed that even though the MNP chains are oriented in the magnetic field direction, their relative distances are so close that the lateral or transversal polarization of the chain assembly in the MNF yields an equivalent dielectric response. A slight difference in perpendicular and parallel permittivity is observed at the lower magnetic volume fraction (1.171%). In this case, the slightly higher parallel permittivity is attributed to the prevailing charge accumulation at the chain terminals (head and tail), which are close to the electrodes. However, for the lowest MNP volume fraction, the two configurations of the fields give the opposite anisotropic effect. Clearly, for perpendicular configuration, the permittivity is slightly higher than the one measured in the parallel configuration. This may reflect in such a structural chain arrangement, in which the transversal polarization of the chains results in a higher bound charge formation in the capacitor than with the one associated with the polarization along the chains. This could happen in the thin MNF sample, with the chains separated from each other enough. Finally, it may be concluded from Figure 5 that the observable permittivity change appears under the action of the 45 mT magnetic field, while the higher field values do not induce further changes in the permittivity of such a thin MNF sample.

3.3. Viscosity

The TO involved in this study exhibits Newtonian behavior with a dynamic viscosity of 15.7 mPa. With the dispersed MNPs of magnetic volume fractions 0.377%, 0.807%, 1.171%, 1.678%, and 2.198%, the viscosity increases to values of 16.53 mPa, 16.76 mPa, 17.3 mPa, 18.31 mPa, and 19.3 mPa, respectively, (at a temperature of 294 K) and also behaves as a Newtonian fluid. The well-known magneto-viscous effect is observed in the presence of the external magnetic field. This is presented in the dynamic viscosity vs. shear rate graphs (flow curves) for the three selected MNF samples in Figure 6.

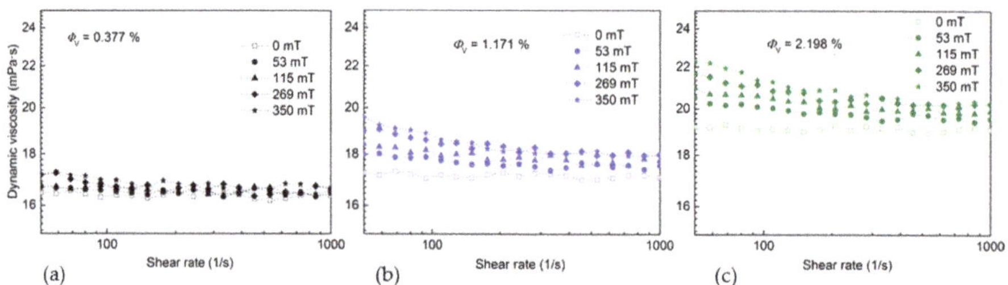

Figure 6. The flow curves under the external magnetic field of various magnitudes plotted for the samples with magnetic volume fractions of 0.377% (**a**), 1.171% (**b**), and 2.198% (**c**).

In Figure 6a, a weak magneto-viscous effect is observed. Even though a quite strong magnetic field (350 mT) is acting on the low-concentrated MNF, the magneto-viscous effect does not arise markedly. This is attributed to the low MNP concentration in which the probability of greater MNP chain formation in the shear flow is low. However, from Figure 6b,c one can see the increasing magneto-viscous effect with the increasing magnetic field. This phenomenon is associated with the magnetic and mechanic torques, which hinder the free rotation of MNPs and their chains in the shear flow. Here, one should realize that the magnetic field is oriented perpendicular to the vorticity of the shear flow. The perpendicular orientation of the MNP chains, which are composed of a small fraction of larger MNPs in the MNF, then leads to an increase in the dynamic viscosity, especially at low shear rates. However, at higher shear rates, the MNP chains disrupt due to intensive fluid flow and this gives rise to the observed shear thinning. Thus, the non-Newtonian behavior with a shear-thinning effect is confirmed for this MNF. The shear-thinning and the magneto-viscous effect become stronger with increasing magnetic volume fraction. A more detailed description of the mechanism and condition leading to the magneto-viscous effect may be found in reference [43].

4. Conclusions

The measurements of thermal conductivity of the transformer oil-based magnetic nanofluid revealed a moderate increase in thermal conductivity with increasing magnetic volume fraction. In the presence of the external magnetic field, a remarkable anisotropy in thermal conductivity was found. The apparently strange anisotropy was discussed in consideration of the magnetic field gradient around the thermal conductivity sensor. The arrangement of the magnetic nanoparticle aggregates in the magnetic field gradient is a decisive factor determining the local thermal conductivity in the magnetic nanofluid. Dielectric spectroscopy measurements of the magnetic nanofluid in the thin sample cell revealed the magnetodielectric anisotropy dependent on the magnetic volume fraction. The studied nanofluid exhibits non-Newtonian behavior with the remarkable shear thinning in the external magnetic field. Based on the obtained results one can conclude that the increased thermal conductivity, low dielectric permittivity, and viscosity make the studied magnetic nanofluid suitable for further testing in real electrical equipment, such as power transformers.

Thus, the investigated parameters of the magnetic nanofluid depend on the magnetic field and the related nanoparticle suspension, aggregate state, or morphology within the bulk liquid. However, the applied experimental methods in this study do not allow one to see the aggregate or agglomeration morphology. Moreover, the three measurement techniques do not use the equivalent magnetic field distributions and sample geometry. This limitation complicates the linking of the three parameters together in a reliable empirical correlation. On the other hand, fitting correlations for these properties would require further comprehensive investigation at various temperatures. Therefore, the presented results might challenge other researchers to carry out simultaneous (in situ) measurements of a particular parameter investigated in this study and the structural properties, e.g., by means of in situ scattering techniques.

Author Contributions: M.R. conceived the study, contributed to the dielectric, magnetic, and thermal measurements and analysis, and wrote the paper. Z.W., B.S. and L.W. designed and carried out the thermal conductivity measurements, B.D. performed the simulations of the magnetic field distribution, and J.K. and R.C. designed the dielectric spectroscopy experiments; K.P. was responsible for the sample preparation and J.T. designed and carried out the measurements of viscosity. P.K. and M.T. were responsible for the resources, formal analysis, and presentation of the study.

Funding: This research was funded by the Slovak Academy of Sciences and Ministry of Education in the framework of the projects: VEGA 2/0141/16, 1/0250/18 and 1/0340/18; the Slovak Research and Development Agency under the Contract Number APVV-18-0160 and APVV-15-0438; the Ministry of Education Agency for structural funds of EU Project ITMS 313011T565, ERANET FMF, ERANET MAGBBRIS and Cultural and Educational Grant Agency of the Ministry of Education, Science, Research and Sport of the Slovak Republic (KEGA) under the project No. 008TUKE-4/2019. APC was funded by Nanouptake COST Action (European Cooperation in Science and Technology), www.cost.eu.

Acknowledgments: This article is based upon work from COST Action Nanouptake, supported by COST (European Cooperation in Science and Technology) www.cost.eu. The study is the result of Nanouptake STSM accomplished by Michal Rajnak and Zan Wu. Zan Wu, Milan Timko, and Peter Kopcansky acknowledge the COST Action Nanouptake for financial support in the participation of the 1st International Conference on Nanofluids (ICNf) and the 2nd European Symposium on Nanofluids (ESNf).

Conflicts of Interest: The authors declare no conflict of interest.

References

1. Fofana, I. 50 years in the development of insulating liquids. *IEEE Electr. Insul. Mag.* **2013**, *29*, 13–25. [CrossRef]

2. Wang, X.; Tang, C.; Huang, B.; Hao, J.; Chen, G. Review of research progress on the electrical properties and modification of mineral insulating oils used in power transformers. *Energies* **2018**, *11*, 487. [CrossRef]
3. Bhunia, M.M.; Panigrahi, K.; Das, S.; Chattopadhyay, K.K.; Chattopadhyay, P. Amorphous graphene—Transformer oil nanofluids with superior thermal and insulating properties. *Carbon* **2018**, *139*, 1010–1019. [CrossRef]
4. Sen Gupta, S.; Manoj Siva, V.; Krishnan, S.; Sreeprasad, T.S.; Singh, P.K.; Pradeep, T.; Das, S.K. Thermal conductivity enhancement of nanofluids containing graphene nanosheets. *J. Appl. Phys.* **2011**, *110*, 084302. [CrossRef]
5. Chen, J.; Sun, P.; Sima, W.; Shao, Q.; Ye, L.; Li, C. A promising nano-insulating-oil for industrial application: Electrical properties and modification mechanism. *Nanomaterials* **2019**, *9*, 788. [CrossRef] [PubMed]
6. Pislaru-Danescu, L.; Morega, A.M.; Telipan, G.; Morega, M.; Dumitru, J.B.; Marinescu, V. Magnetic nanofluid applications in electrical engineering. *IEEE Trans. Magn.* **2013**, *49*, 5489–5497. [CrossRef]
7. Rajnak, M.; Timko, M.; Kopcansky, P.; Paulovicova, K.; Kuchta, J.; Franko, M.; Kurimsky, J.; Dolnik, B.; Cimbala, R. Transformer oil-based magnetic nanofluid with high dielectric losses tested for cooling of a model transformer. *IEEE Trans. Dielectr. Electr. Insul.* **2019**, *26*, 1343–1349. [CrossRef]
8. Nkurikiyimfura, I.; Wang, Y.; Pan, Z. Heat transfer enhancement by magnetic nanofluids—A review. *Renew. Sustain. Energy Rev.* **2013**, *21*, 548–561. [CrossRef]
9. Lee, M.; Kim, Y.-J. Thermomagnetic Convection of Ferrofluid in an Enclosure Channel with an Internal Magnetic Field. *Micromachines* **2019**, *10*, 553. [CrossRef]
10. Vatani, A.; Woodfield, P.L.; Nguyen, N.T.; Dao, D.V. Onset of thermomagnetic convection around a vertically oriented hot-wire in ferrofluid. *J. Magn. Magn. Mater.* **2018**, *456*, 300–306. [CrossRef]
11. Bianco, V.; Manca, O.; Nardini, S.; Vafai, K. *Heat Transfer Enhancement with Nanofluids*; CRC Press: Boca Raton, FL, USA, 2015; ISBN 1482254026.
12. Doganay, S.; Alsangur, R.; Turgut, A. Effect of external magnetic field on thermal conductivity and viscosity of magnetic nanofluids: A review. *Mater. Res. Express* **2019**, *6*, 112003. [CrossRef]
13. Patel, J.; Parekh, K.; Upadhyay, R.V. Maneuvering thermal conductivity of magnetic nanofluids by tunable magnetic fields. *J. Appl. Phys.* **2015**, *117*, 243906. [CrossRef]
14. Philip, J.; Shima, P.D.; Raj, B. Enhancement of thermal conductivity in magnetite based nanofluid due to chainlike structures. *Appl. Phys. Lett.* **2007**, *91*, 203108.
15. Shima, P.D.; Philip, J.; Raj, B. Magnetically controllable nanofluid with tunable thermal conductivity and viscosity. *Appl. Phys. Lett.* **2009**, *95*, 133112. [CrossRef]
16. Altan, C.L.; Elkatmis, A.; Yüksel, M.; Aslan, N.; Bucak, S. Enhancement of thermal conductivity upon application of magnetic field to Fe_3O_4 nanofluids. *J. Appl. Phys.* **2011**, *110*, 093917.
17. Parekh, K.; Lee, H.S. Magnetic field induced enhancement in thermal conductivity of magnetite nanofluid. *J. Appl. Phys.* **2010**, *107*, 09A310.
18. Wu, Z.; Sundén, B. Convective heat transfer performance of aggregate-laden nanofluids. *Int. J. Heat Mass Transf.* **2016**, *93*, 1107–1115. [CrossRef]
19. Odenbach, S. *Colloidal Magnetic Fluids: Basics, Development and Application of Ferrofluids*; Springer: Berlin, Germany, 2009; ISBN 9783540853862.
20. Jin, J.; Song, D.; Geng, J.; Jing, D. Time-dependent scattering of incident light of various wavelengths in ferrofluids under external magnetic field. *J. Magn. Magn. Mater.* **2018**, *447*, 124–133. [CrossRef]
21. Gavili, A.; Zabihi, F.; Isfahani, T.D.; Sabbaghzadeh, J. The thermal conductivity of water base ferrofluids under magnetic field. *Exp. Therm. Fluid Sci.* **2012**, *41*, 94–98. [CrossRef]
22. Katiyar, A.; Dhar, P.; Nandi, T.; Das, S.K. Magnetic field induced augmented thermal conduction phenomenon in magneto-nanocolloids. *J. Magn. Magn. Mater.* **2016**, *419*, 588–599. [CrossRef]
23. Li, Q.; Xuan, Y.; Wang, J. Experimental investigations on transport properties of magnetic fluids. *Exp. Therm. Fluid Sci.* **2005**, *30*, 109–116. [CrossRef]
24. Segal, V.; Hjortsberg, A.; Rabinovich, A., Nattrass, D.; Raj, K. AC (60 Hz) and impulse breakdown strength of a colloidal fluid based on transformer oil and magnetite nanoparticles. In Proceedings of the Conference Record of the 1998 IEEE International Symposium on Electrical Insulation (Cat. No.98CH36239), Arlington, VA, USA, 7–10 June 1998; Volume 2, pp. 619–622.
25. Lee, J.-C.; Seo, H.-S.; Kim, Y.-J. The increased dielectric breakdown voltage of transformer oil-based nanofluids by an external magnetic field. *Int. J. Therm. Sci.* **2012**, *62*, 29–33. [CrossRef]
26. Kurimský, J.; Rajňák, M.; Cimbala, R.; Rajnič, J.; Timko, M.; Kopčanský, P. Effect of magnetic nanoparticles on partial discharges in transformer oil. *J. Magn. Magn. Mater.* **2019**, *496*, 165923. [CrossRef]

27. Kurimský, J.; Rajňák, M.; Bartko, P.; Paulovičová, K.; Cimbala, R.; Medved, D.; Džamová, M.; Timko, M.; Kopčanský, P. Experimental study of AC breakdown strength in ferrofluid during thermal aging. *J. Magn. Magn. Mater.* **2018**, *465*, 136–142. [CrossRef]
28. Hwang, J.G.; Zahn, M.; O'Sullivan, F.M.; Pettersson, L.A.A.; Hjortstam, O.; Liu, R. Effects of nanoparticle charging on streamer development in transformer oil-based nanofluids. *J. Appl. Phys.* **2010**, *107*, 014310. [CrossRef]
29. Rajnak, M.; Spitalsky, Z.; Dolnik, B.; Kurimsky, J.; Tomco, L.; Cimbala, R.; Kopcansky, P.; Timko, M. Toward Apparent Negative Permittivity Measurement in a Magnetic Nanofluid with Electrically Induced Clusters. *Phys. Rev. Appl.* **2019**, *11*, 024032. [CrossRef]
30. Rajnak, M.; Kurimsky, J.; Dolnik, B.; Kopcansky, P.; Tomasovicova, N.; Taculescu-Moaca, E.A.; Timko, M. Dielectric-spectroscopy approach to ferrofluid nanoparticle clustering induced by an external electric field. *Phys. Rev. E* **2014**, *90*, 032310. [CrossRef]
31. Malaescu, I.; Marin, C.N. Dependence on the temperature of the activation energy in the dielectric relaxation processes for ferrofluids in low-frequency field. *J. Magn. Magn. Mater.* **2002**, *252*, 68–70. [CrossRef]
32. Spanoudaki, A.; Pelster, R. Frequency dependence of dielectric anisotropy in ferrofluids. *J. Magn. Magn. Mater.* **2002**, *252*, 71–73. [CrossRef]
33. Shahrivar, K.; Morillas, J.R.; Luengo, Y.; Gavilan, H.; Morales, P.; Bierwisch, C.; de Vicente, J. Rheological behavior of magnetic colloids in the borderline between ferrofluids and magnetorheological fluids. *J. Rheol.* **2019**, *63*, 547–558. [CrossRef]
34. Rosa, A.P.; Cunha, F.R. The influence of dipolar particle interactions on the magnetization and the rotational viscosity of ferrofluids. *Phys. Fluids* **2019**, *31*, 052006. [CrossRef]
35. Susan-Resiga, D.; Socoliuc, V.; Boros, T.; Borbáth, T.; Marinica, O.; Han, A.; Vékás, L. The influence of particle clustering on the rheological properties of highly concentrated magnetic nanofluids. *J. Colloid Interface Sci.* **2012**, *373*, 110–115. [CrossRef] [PubMed]
36. Odenbach, S.; Pop, L.M.; Zubarev, A.Y. Rheological properties of magnetic fluids and their microstructural background. *GAMM Mitt.* **2007**, *30*, 195–204. [CrossRef]
37. Shojaeizadeh, E.; Veysi, F.; Goudarzi, K.; Feyzi, M. Magnetoviscous effect investigation of water based Mn-$ZnFe_2O_4$ magnetic nanofluid under the influence of magnetic field: An experimental study. *J. Magn. Magn. Mater.* **2019**, *477*, 292–306. [CrossRef]
38. Vékás, L.; Bica, D.; Avdeev, M.V. Magnetic nanoparticles and concentrated magnetic nanofluids: Synthesis, properties and some applications. *China Particuol.* **2007**, *5*, 43–49. [CrossRef]
39. Rosensweig, R.E. *Ferrohydrodynamics*; Dover Publications: New York, NY, USA, 2013; ISBN 0486783006.
40. Prasher, R.; Evans, W.; Meakin, P.; Fish, J.; Phelan, P.; Keblinski, P. Effect of aggregation on thermal conduction in colloidal nanofluids. *Appl. Phys. Lett.* **2006**, *89*, 143119. [CrossRef]
41. Evans, W.; Prasher, R.; Fish, J.; Meakin, P.; Phelan, P.; Keblinski, P. Effect of aggregation and interfacial thermal resistance on thermal conductivity of nanocomposites and colloidal nanofluids. *Int. J. Heat Mass Transf.* **2008**, *51*, 1431–1438. [CrossRef]
42. Rajnak, M.; Dolnik, B.; Kurimsky, J.; Cimbala, R.; Kopcansky, P.; Timko, M. Electrode polarization and unusual magnetodielectric effect in a transformer oil-based magnetic nanofluid thin layer. *J. Chem. Phys.* **2017**, *146*, 014704. [CrossRef]
43. Odenbach, S.; Thurm, S. Magnetoviscous Effects in Ferrofluids. In *Ferrofluids*; Springer: Berlin/Heidelberg, Germany, 2002; pp. 185–201.

Conference Report

From Thermal to Electroactive Graphene Nanofluids †

Daniel Rueda-García, María del Rocío Rodríguez-Laguna, Emigdio Chávez-Angel, Deepak P. Dubal, Zahilia Cabán-Huertas, Raúl Benages-Vilau and Pedro Gómez-Romero *

Catalan Institute of Nanoscience and Nanotechnology, ICN2 (CSIC-BIST), Campus de la UAB, 08193 Bellaterra (Barcelona), Spain; ruedizzi@gmail.com (D.R.-G.); rodriguez3laguna@gmail.com (M.d.R.L.); emigdio.chavez@icn2.cat (E.C.-A.); dubaldeepak2@gmail.com (D.P.D.); hazalia@gmail.com (Z.C.-H.); raul.benages@icn2.cat (R.B.-V.)

* Correspondence: pedro.gomez@icn2.cat

† This paper is an extended version of our paper published in ICNf2019. 1st International Conference on Nanofluids, Castellón, Spain, 26–28 June 2019.

Received: 8 November 2019; Accepted: 26 November 2019; Published: 28 November 2019

Abstract: Here, we describe selected work on the development and study of nanofluids based on graphene and reduced graphene oxide both in aqueous and organic electrolytes. A thorough study of thermal properties of graphene in amide organic solvents (N,N-dimethylformamide, N,N-dimethylacetamide, and N-methyl-2-pyrrolidone) showed a substantial increase of thermal conductivity and specific heat upon graphene integration in those solvents. In addition to these thermal studies, our group has also pioneered a distinct line of work on electroactive nanofluids for energy storage. In this case, reduced graphene oxide (rGO) nanofluids in aqueous electrolytes were studied and characterized by cyclic voltammetry and charge-discharge cycles (i.e., in new flow cells). In addition, hybrid configurations (both hybrid nanofluid materials and hybrid cells combining faradaic and capacitive activities) were studied and are summarized here.

Keywords: graphene; reduced graphene oxide (rGO); nanofluids; thermal properties; heat transfer fluids; electrochemical energy storage; electroactive nanofluids

---

1. Introduction

An increasing number of applications related to energy conversion and storage rely on graphene because of its extraordinary combination of properties [1,2]. Graphene is a solid material and it has been used as such in all these applications, however, fluids are strategic materials used in a wide range of industrial applications, which span from thermal to biomedical or to electrochemical systems. In particular, nanofluids, which integrate solid nanoparticles dispersed in a base liquid and constitute a new type of materials with ground-breaking new properties, provide new opportunities to advance in many fields. Heat transfer is currently the most intensively explored application. However, magnetic ferrofluids, health applications, and energy storage appear as other promising fields of study and potential application [3,4].

The nature of the solid phases used in the preparation of nanofluids is extremely varied. In the case of heat transfer fluids (HTFs), all types of solids, from metals to oxides to carbons have been widely studied given the superior thermal conductivity of solids as compared to liquids [5], however, magnetic or electrochemical nanofluids are much more restricted to phases with the necessary magnetic or electroactive nature. In the latter type, electroactivity can be redox [6,7] or capacitive [8], although hybrid materials and devices combining both of those are also possible [9]. In electroactive nanofluids, nanoparticles are dispersed into a base fluid that must be an ionic-conducting electrolyte. This represents an additional challenge in order to avoid coagulation processes which are frequently associated with the presence of ionic salts in the medium.

Electroactive nanofluids can be used in flow cells. These are a special type of energy storage system in which electroactive fluids are stored out of the electrochemical cell and are forced to circulate through it by pumping. This configuration leads to the decoupling of energy density (which is determined by the volume of the external tanks) and power (which will depend on the active area of the cell electrodes). This has led to the development of modular designs that are best suited for stationary energy storage, but able to meet any requirement, and therefore are a perfect solution to be coupled with intermittent power generation similar to that associated with renewable energies.

Graphene nanofluids are prepared by dispersing graphene (or RGO) nanosheets in an adequate base fluid. They can be stabilized in organic or aqueous solvents [6,8,10,11] in the form of pure, non-oxidized graphene [11] or rGO [8], but also in the form of hybrids [6,10]. All of these solvents present useful thermal or electrochemical properties, among others, as we summarize below. Indeed, this study is an extended overview of our recent work on functional nanofluids with an emphasis on their thermal and especially on their electrochemical properties and applications, as presented at the 1st International Conference of Nanofluids [12].

2. Materials and Methods

Graphene flakes with lateral sizes ~150–450 nm and thicknesses from 1 to 10 layers were prepared from graphite (Sigma-Aldrich, purity >99% and size <20 μm) by a mechanical exfoliation method, as described in a previous work [11].

Graphene oxide (GO) was synthesized from graphite using a modified Hummers method. Briefly, 5 g $NaNO_3$ and 225 ml H_2SO_4 were added to 5 g graphite and stirred for 30 min in an ice bath. Then, 25 g $KMnO_4$ was added to the resulting solution, and the solution was stirred at 50 °C for 2 h. Then, 500 ml deionized water and 30 ml H_2O_2 (35%) were slowly added to the solution, and the solution was washed with dilute HCl. Next, the GO product was washed again with 500 ml concentrated HCl (37%). The reduced graphene oxide was prepared by high temperature treatment of the GO sample at 800 °C.

In the second step, we prepared hybrid materials based on rGO, phosphomolybdic acid (rGO-PMo_{12}, where PMo_{12} is the short notation for $H_3PMo_{12}O_{40}$) and phosphotungstic acid (rGO-PW_{12}, where PW_{12} is the abbreviation for $H_3PW_{12}O_{40}$). Briefly, 0.25 g of rGO was dispersed in 100 ml of deionized water with a probe sonicator (of power 1500 watt) for 1 h. Then, 10 mM of the corresponding POM was added to the presonicated rGO dispersion. This suspension was further sonicated (ultrasonic bath 200 watt) for 5 h and kept at room temperature for 24 h. Afterwards, the product was filtered off and dried in a vacuum oven at 80 °C, overnight.

Electroactive nanofluids were prepared by direct mixing of rGO with the base fluid. In this study, the base fluid is distilled water 1 M H_2SO_4. Nanofluids with different concentrations were prepared by mixing 0.025, 0.1, and 0.4 wt% of rGO in 1 M H_2SO_4 aqueous solution. In order to get a stable dispersion, 0.5 wt% of surfactant (triton X-100) was added and the mixture was kept in an ultrasonic bath for up to 2 h.

Thermal conductivity was measured using a modified three-omega (3ω) method based on the work of Oh et al. [13]. Finally, the stability of the dispersions was studied over time by regular tests every month for four months using dynamic light scattering (DLS) with a ZetaSizer nano ZS (ZEN3600, Malvern Instruments, Ltd., Malvern, UK). The rheological measurements were carried out with a rheometer (HAAKE RheoStress RS600, Thermo Electron Corporation). Electrochemical characterizations such as cyclic voltammetry (CV), galvanostatic charge/discharge (chronopotentiometry, CP), chronoamperometry (CA), and electrochemical impedance measurements of rGO nanofluid symmetric cell were carried out with Biologic SP200 and VMP3 potentiostats.

3. Thermal Graphene Nanofluids

We have designed and prepared stable graphene nanofluids for thermal applications. Several organic solvents were used, in particular N,N-dimethylformamide (DMF),

N,N-dimethylacetamide (DMAc), and N-methyl-2-pyrrolidone (NMP). Thermal conductivity data can be found in Figure 1, which allows comparison of pure DMAc and graphene nanofluids in DMAc (0.09 wt% and 0.18 wt%). A 3-ω method adapted to liquid samples was used to carry out the measurements [11]. Specific heat, thermal conductivity and sound velocity were all found to increase with the content of graphene [11]. Thus, a nanofluid with just 0.18 wt% of graphene yielded an enhancement of 48% in thermal conductivity as compared with the solvent, exemplifying how a very small amount of graphene leads to a very substantial effect.

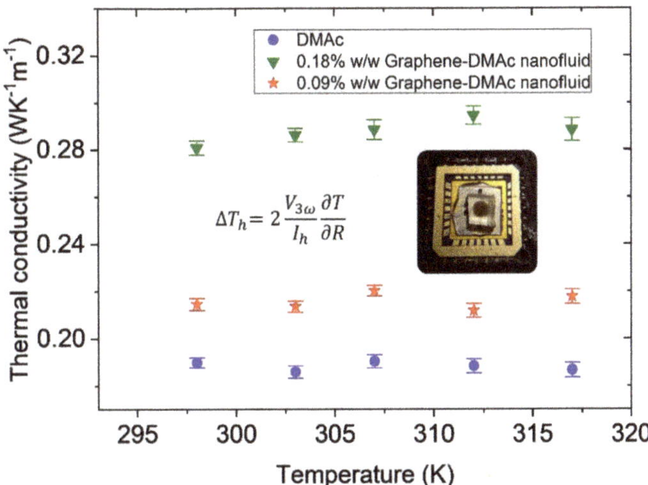

Figure 1. Thermal conductivity as a function of temperature measured through the 3-omega technique for graphene nanofluids in dimethylacetamide solvent.

In order to compare this large effect with theory, we calculated the predicted thermal conductivity enhancement given by effective medium theory (EMT) using the Maxwell equation, given by:

$$\frac{\kappa_m}{\kappa_f} = \frac{\left(2\phi\left(\kappa_p - \kappa_f\right) + 2\kappa_f + \kappa_p\right)}{\left(-\phi\left(\kappa_p - \kappa_f\right) + 2\kappa_f + \kappa_p\right)} \tag{1}$$

where κ_m, κ_f, and κ_p are the thermal conductivities of the effective medium (i.e., nanofluid), base fluid, and nanoparticles, respectively, and ϕ is the particle volume fraction.

If we consider that the $\kappa_f << \kappa_p$ (for DMAc and DMF $\kappa_f = 0.175$ and 0.180 W K^{-1} m^{-1}, respectively, while for graphene $300 < \kappa_p < 3000$ W K^{-1} m^{-1} in the limit of small particle concentrations, Equation (1) is rewritten as:

$$\frac{\kappa_m}{\kappa_f} \cong \frac{(1 + 2\phi)}{(1 - \phi)} \approx (1 + 2\phi)(1 + \phi) \approx 1 + 3\phi \tag{2}$$

Figure 2 (left) shows the comparison using EMT model using Maxwell equation, Equation (1), and a thermal conductivity value of nanoparticles of kp = 600 W K^{-1} m^{-1} and its approximation, Equation (2). It can be observed that for very low concentration values both Equations (1) and (2) do not match our experimental observations. Even for a very large span of kp (Figure 2, right) the enhancement predicted by Equation (1) is negligible.

The very poor matching can be related to the assumptions made within the Maxwell model, which only takes into account thermal conductivities of the base fluid and particles and volume fraction of particles, while particle size, shape, and the distribution and motion of dispersed particles, which are not taken into consideration in this model, may have an unaccounted impact on thermal conductivity enhancement. Therefore, the experimental results could not be compared with the correlated values of this theoretical framework.

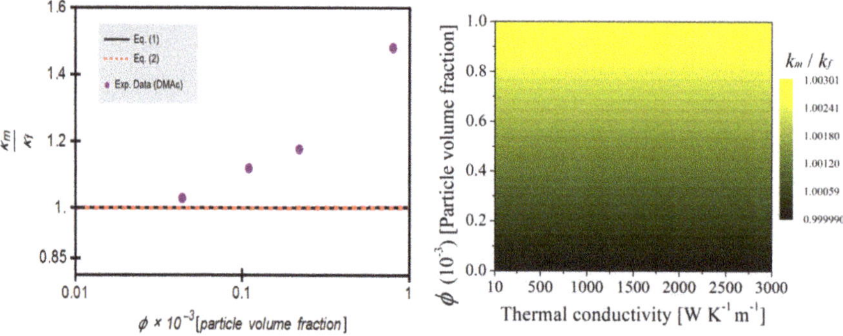

Figure 2. Left: Maxwell equation (Equation (1)) and its approximation (Equation (2)) compared with our measured data. **Right**: Counterplot of Maxwell equation as a function of volume fraction and *kp*.

In this context, a number of works have been published showing different expressions for the effective thermal conductivity of the nanofluids [14]. The modifications to the classical EMT have been mainly associated to the following: particle size, shape, volume fraction, temperature, agglomeration, percolation, nanolayering, and static and dynamic conditions of nanoparticles (Brownian motion and localized convection). Among this sea of theoretical models, one widely used for graphene flakes and nanowires is the so-called Nan's model [15]. According to Nan's model, the thermal conductivity of the nanofluid can be calculated as follows:

$$\frac{\kappa_m}{\kappa_f} = \frac{\phi\,(2\beta_{11}\,(1 - L_{11}) + \beta_{33}\,(1 - L_{33})) + 3}{(3 - \phi\,(2\beta_{11}L_{11} + \beta_{33}L_{33}))} \tag{3}$$

$$\beta_{ii} = \frac{\kappa_p - \kappa_f}{L_{ii}\left(\kappa_p - \kappa_f\right) + \kappa_f} \tag{4}$$

where L_{ii} is a form factor which depends on the geometry of the nanoparticle. For the case of graphene flakes, the aspect ratio is very high, and therefore $L_{11} = 0$ and $L_{33} = 1$ [16]. Figure 3 compares Nan's model calculations with our experimental data showing a much better correlation with this model.

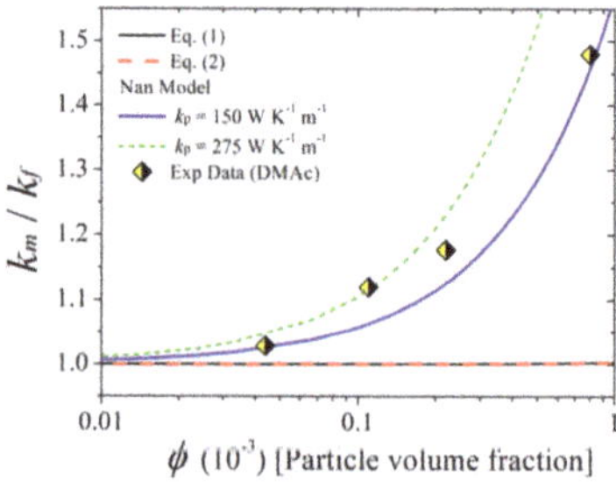

Figure 3. Nan's model and its comparison with our measured data.

In addition to the described trend of thermal conductivity enhancement in graphene nanofluids, we detected a very interesting correlation of this effect (for DMF and DMAc nanofluids) with an overall shift of specific Raman bands in the series of nanofluids (see Figure 4). No band splitting is observed. Instead, the whole bands are broadened and clearly shifted to higher frequencies, with the shift being proportional to the concentration of graphene. This observation implies that the effect of adding graphene affects the bulk solvent. It should be noted that this shift only takes place for solvents in which the thermal conductivity is increased by graphene, in our case DMF and DMAc, whereas for NMP nanofluids neither a significant thermal conductivity enhancement nor Raman shifts were observed [17]. Possible mechanisms explaining this behavior have been proposed and analyzed elsewhere [11,17].

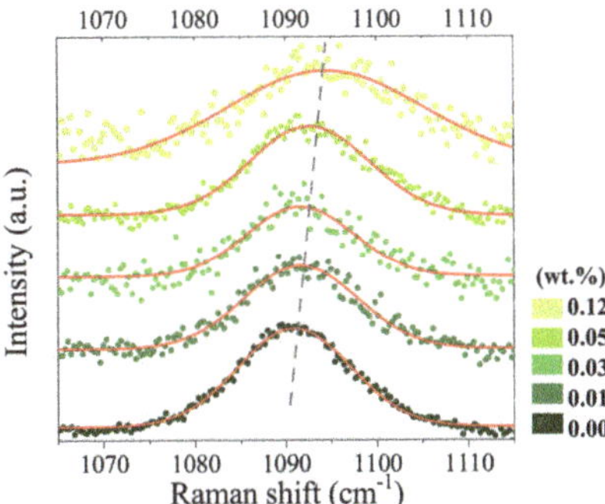

Figure 4. Raman spectra of pure N,N-dimethylformamide (DMF) (**bottom**) and graphene nanofluids with increasing concentrations showing the shift and broadening of the rocking mode peak ascribed to the "(CH3)N" bond in DMF (from ref [11]).

4. Electroactive Graphene Nanofluids

In the area of electrochemical energy storage, our group has pioneered the design and study of electroactive graphene nanofluids. In this case, rGO (prepared by Hummers method, then thermally reduced at 800 °C) was dispersed in acidic aqueous electrolytes and we were able to demonstrate a very fast charge transfer [8,10]. Thus, a dispersion of rGO in 1M H_2SO_4 stabilized in aqueous solution with the addition of a surfactant (0.5 wt% of triton X-100 and sonicated by ultrasonic bath up to 2 h), featured low viscosity values, close to those of water (in the shear rate range of 25 to 150 per s at room temperature). Dispersions of graphene in organic solvents, as the ones described for thermal nanofluids, could also be used for the development of electroactive nanofluids, provided a suitable salt could be dissolved to form the necessary electrolytic solution. Organic solvents would be best suited for (non-oxidized) graphene materials, whereas aqueous electrolytes are perfect to disperse GO or rGO. In our case, preliminary tests, with the addition of various salts to pure graphene in organic solvents, led to dispersions which were less stable in time than those of rGO in dilute sulfuric acid, which is why we proceeded with the latter as a case study for electroactive nanofluids. The specific capacity of this dispersion was similar to that found for solid electrodes in conventional rGO supercapacitors (169 F/g(rGO)). Remarkably, these nanofluids were able to work at faster rates, up to 10,000 mV/s, in 0.025, 0.1, and 0.4 wt% rGO (Figure 5). This fast response confirms the great potential of these nanofluid materials for applications in novel devices which we could call flowing supercapacitors. We should note that even very dilute nanofluids led to full utilization of the dispersed active material.

This effective charge-transfer capacity was corroborated by studying the nanofluid performance under continuous flow conditions. Under flowing conditions, the CVs show identical profiles and they remain identical as flow rates are changed (Figure 6, LEFT). Moreover, we found that the specific capacitance improves under flow conditions (Figure 6 RIGHT) in contrast to semisolid viscous slurries, for which the conductivity, and therefore the capacitance, decrease notably under flow conditions [18]. We found the maximum capacitance at a flow rate of 10 ml/min in our homemade serpentine flow channel cell.

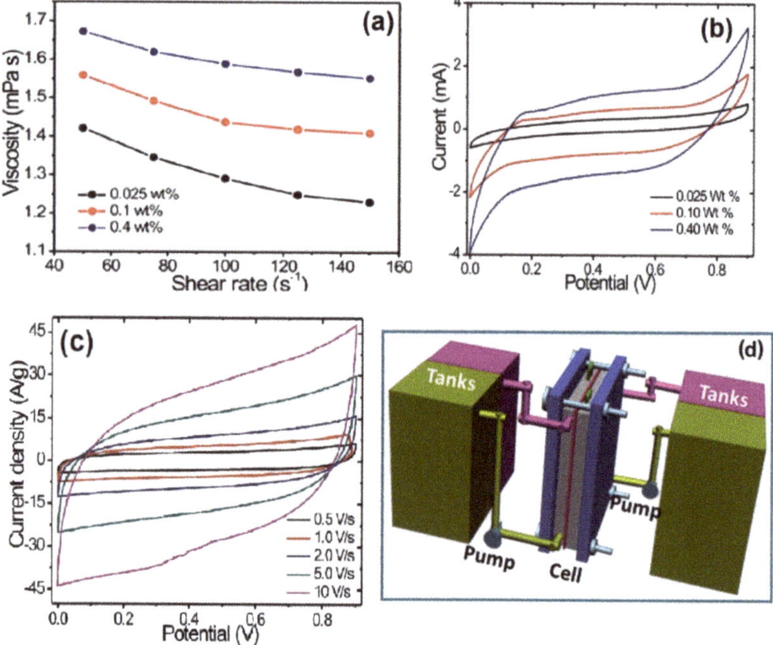

Figure 5. (**a**) Rheological properties of the three nanofluids studied showing low viscosity suitable for efficient pumping, cyclic voltammetry (CV) curves of (**b**) rGO nanofluids of different concentration at 20 mV/s scan rate in static condition and (**c**) rGO nanofluid of 0.025 wt% concentration at different scan rates (data from ref. [8]) (**d**) scheme of the electrochemical system used.

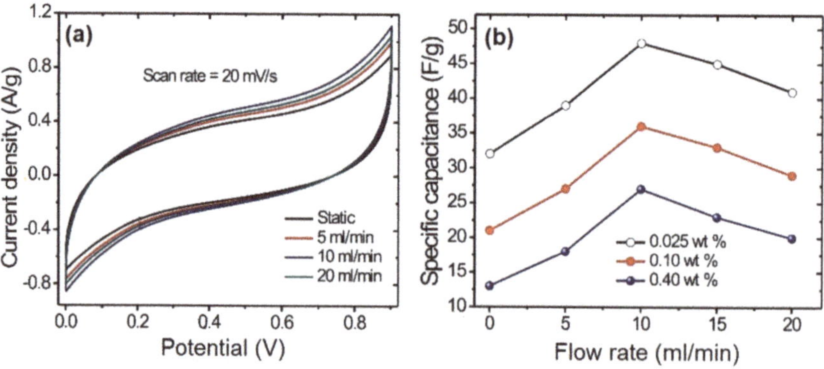

Figure 6. LEFT: CVs (20 mV/s) of 0.025 wt% rGO nanofluids at various flow rates. **RIGHT**: Specific capacitance of varios rGO nanofluids as a function of flow rate (data from ref. [8]).

Concerning the viscosity of our nanofluids, we tried to compare our experimental results with theoretical models (Einstein-Batchelor) but the results showed a poor agreement between our data and the model. However, this is not too surprising given the strong assumptions made in that model, especially the spherical nature of the nanoparticle modeled as compared with the strongly anisotropic nature of graphene nanosheets.

We have also studied nanofluids based on hybrids which in turn were made combining graphene as a conductive and capacitive material with redox-active inorganic species. Thus, some of the hybrid electroactive nanofluids (HENFs) that we have prepared integrate faradaic and capacitive storage mechanisms simultaneously in a nanofluid. Specifically, we have designed and prepared "liquid electrodes" based on a stable nanocomposite of rGO and polyoxometalates (POMs) and have studied them in a novel flow cell that differs from conventional ones precisely in the use of nanofluids. Two rGO nanocomposites were prepared and a surfactant was used to disperse them in 1M H_2SO_4. The hybrids used two different polyoxometalates (POMs), phosphomolybdic acid (rGO-PMo_{12}) and phosphotungstic acid (rGO-PW_{12}), which were synthesized by mixing rGO with a solution of the corresponding POM as previously reported [19].

We used a homemade flow cell system (carved serpentine flow 200 mm long path, 5 mm wide, and 1mm in depth, with total cell size of 7 cm × 6 cm × 1 cm) to study these hybrid electroactive nanofluids (HENFs) both under flowing and static conditions [10]. Both nanofluids showed a viscosity very close to that of water (Figure 7a). It should be pointed out that even at low concentrations of rGO-POM (0.025 wt%) high values of specific capacitance were measured for rGO-PW_{12} (273 F/g) and for rGO-PMo_{12} (305 F/g) leading to high specific energy and power densities (Figure 7). Furthermore, after 2000 cycles a good coulombic efficiency (~77% to 79%) was reached and an excellent cycling stability (~95%). According to these results, our POM HENFs act as genuine liquid electrodes and feature outstanding properties. This supports the potential application of this type of HENFs for energy storage applications in nanofluids flow cells.

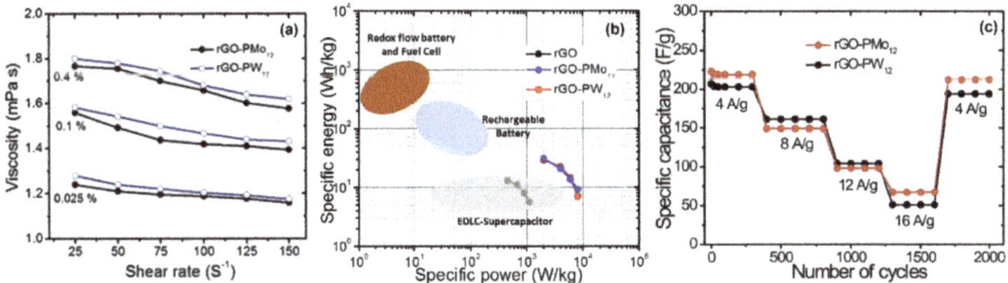

Figure 7. (**a**) The viscosity of rGO, rGO-PMo12 and rGO-PW12 HENFs of different concentrations with shear rate. (**b**) Ragone plot for rGO, rGO-PW12 and rGO-PMo12 nanofluids described in this paper and comparison with other energy storage technologies [9,20–23]. (c) Galvanostatic cycling performance for rGO-POMs hybrid electroactive nanofluids of 0.025% at various current densities for 2000 cycles (data from ref [10]).

It should be underscored that we were able to fully utilize the dispersed rGO nanosheets. Both our energy and power density values are even slightly better (6 to 30 Wh $kg{-}1_{rGO\text{-}POM}$ and 2 to 9 kW $kg{-}1_{rGO\text{-}POM}$) than those reported in the literature for the corresponding solid electrodes [24]. On the one hand, it should also be noted that in our case only a small fraction of the fluid mass is electroactive (in contrast with conventional solid electrodes). Indeed, increasing the load of active material is the most important remaining challenge for these novel electroactive nanofluids. A greater concentration of electroactive rGO or a greater concentration of POM in rGO-POM would lead to a proportionally greater capability to store electrons per unit volume of bulk fluid. On the other hand, both aqueous reduced graphene nanofluids showed limited stability. Ten hours after dispersion a change in the opacity of the nanofluids could be appreciated due to the precipitation of rGO, and after 40 hours almost all the rGO had precipitated.

We developed a new kind of aqueous nanofluid as proof of concept to obtain a solution with increased stability of the rGO without using any surfactant and at the same time keeping the fast charge transfer capability and low viscosity. Thus, in order to improve the stability of rGO, we dissolved an aromatic molecule in the dispersion with the potential to establish π–π interactions with rGO, avoiding or at least hindering its restacking and precipitation, but at the same time leaving free the edges to transfer the charge between rGO layers without increasing the viscosity. We used 3,4-diaminobenzoic acid (DABA) as the dissolved aromatic molecule to stabilize rGO. We found that with a 40:1 mass ratio of DABA to rGO (0.3g/L) the nanofluid is stable up to 97 h instead of 10 h, which is a very substantial improvement [6]. This nanofluid also had LiOH to adjust the pH to seven to ensure good solubility of DABA and as electrolyte salt.

$LiFePO_4$ was chosen as a typical battery material to test the charge transfer capacity of our DABA-stabilized rGO nanofluid. Thus, various amounts of $LiFePO_4$ nanoparticles were added to the DABA/rGO nanofluid. On the one hand, it should be noted that neither rGO alone, nor $LiFePO_4$ nanoparticles alone (both dispersed in the same LiOH/1M Li_2SO_4 electrolyte), showed any redox response (Figure 8).

On the other hand, $LiFePO_4$ nanoparticles dispersed in the DABA/rGO nanofluid showed the characteristic redox peaks corresponding to Li+ de- and intercalation in $LiFePO_4$. These redox peaks were clearly observed even at fast scan rates (for an intercalation material) of 25 mV/s (Figure 9), indeed, we were able to detect the redox peaks at 100mV/s, but above 25 mV/s the separation between them increased, implying an excessive resistance at those scan rates.

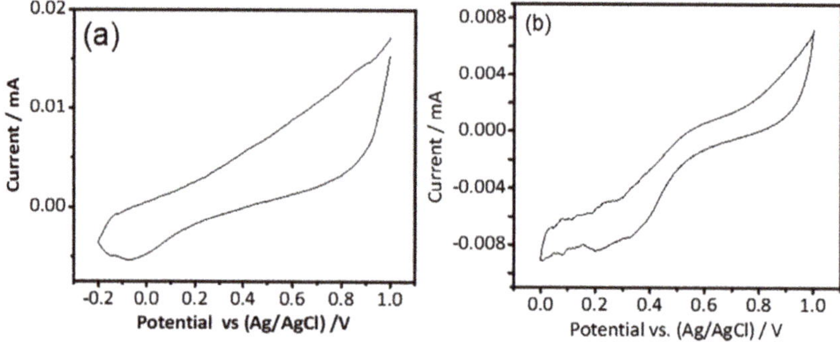

Figure 8. CVs of (**a**) reduced graphene oxide (rGO) and (**b**) $LiFePO_4$ (1.4 g/L) in water electrolyte (scan rate 5 mV/s). (Reprinted with permission from Rueda-García, Daniel et al. This article was published in Electrochemica Acta, 281, 598 (2018), copyright 2018 Elsevier.)

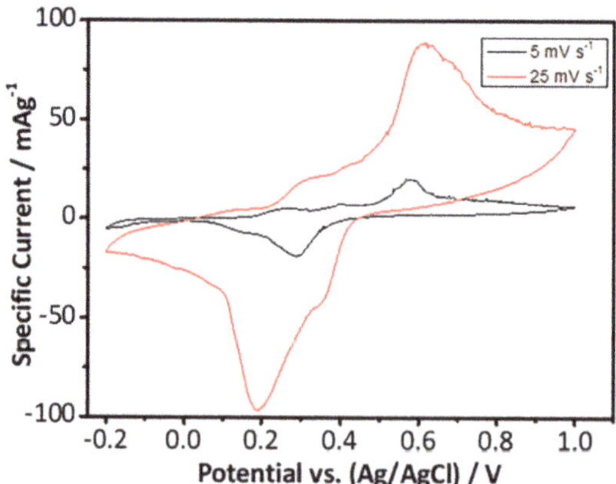

Figure 9. CV of faradaic $LiFePO_4$ (1 g/L) with capacitive3,4-diaminobenzoic acid reduced graphene oxide (DABA/rGO) (40:1). (Reprinted with permission from Rueda-García, Daniel et al. This article was published in Electrochemica Acta, 281, 598 (2018), copyright 2018 Elsevier.)

The detection of $LiFePO_4$ redox peaks only in the presence of rGO proves that rGO provides the electrons needed for the redox process of the $LiFePO_4$ nanoparticles in an effective way that the inorganic nanoparticles could not get by themselves due to their sluggish diffusion in the current collectors. Moreover, we were also able to fully charge and discharge the $LiFePO_4$ nanoparticles dispersed with high coulombic efficiency which proves that rGO provides charge to the total volume of the nanofluid, thus, leading to a full charge and discharge of the bulk material in the nanofluid.

5. Summary and Conclusions

Notwithstanding the importance of nanofluids for thermal energy transfer and storage, which we have also studied, we have shown in this article the potential of nanofluids in other novel applications, namely, electrochemical energy storage. From the results shown here, we can induce important general conclusions concerning the extended effects of small amounts of solid (in our case graphene) throughout the whole volume of the nanofluid. These effects are present both in thermal, as well as in electrochemical systems. Thus, thermal conductivity is notably enhanced (48%) by a tiny amount of graphene (0.18%) because, surprisingly, this tiny amount of graphene affects the bulk of the solvent molecules. Furthermore, other properties such as specific heat and sound velocity also change remarkable, and we have been able to show a correlation between the enhancement of these thermal properties and the Raman spectra of the bulk fluids.

Nevertheless, we have proven that stable rGO water dispersions are able to transfer charge through all the nanofluid volume, which lead to the whole nanofluid acting as a supercapacitor electrode storing charge through a capacitive mechanism. Indeed, the rGO aqueous nanofluid showed an extremely fast charge transfer, capable of cycling at 10V/s. Thanks to this fast charge transfer we were able to completely charge, and discharge, dispersed redox active nanoparticles of $LiFePO_4$ and clearly detect its redox waves even at 25 mV/s. Moreover, doping the rGO with molecular active redox species, such as the polyoxometalates, we developed hybrid systems with improved power and capacity with respect to the pure rGO nanofluid.

Finally, we have shown the outstanding stability of graphene nanofluids in amide solvents (DMF, DMAc, and NMP) in the absence of any surfactant and have shown how to stabilize rGO in aqueous saline electrolytes (we proved that aqueous rGO nanofluids can improve their stability by dissolving an aromatic molecule (DABA) able to stabilize rGO by π–π interactions while keeping its good electrical conductivity). All of this has been possible maintaining the viscosity of the nanofluids developed very close to that of the parent solvents, which will facilitate their final application in real flowing devices. However, the low concentration of graphene nanosheets could be a handicap for the application of these materials in high-energy density devices. Therefore, an important goal is increasing the loading of electroactive nanoparticles. Other pioneering groups, for example, Timofeeva et al. have been able to show that it is possible to develop electroactive nanofluids which can maintain low viscosity and high loading of electroactive nanoparticles (surface-modified TiO_2 in their case) [7,25], thus, showing that this is not an intrinsic barrier for this type of material.

Furthermore, other groups are also contributing to the development of this field, working both with polyoxometalate clusters [26] or oxides (Fe_2O_3) [27] and exploring new media such as ionic liquids as base fluid for electroactive nanofluids [27].

In summary, we have designed and prepared nanofluids based on graphene but also on graphene hybrids, both in aqueous and organic electrolytes. We have shown how these novel nanofluid materials can feature outstanding performances even in the case of very dilute systems. Both, in the case of thermal properties as well as when it comes to electrochemical properties, we have demonstrated nonlinear effects, leading to remarkable properties with small amounts of graphene dispersed in the nanofluids. Thus, our work underscores the solid potential of these systems as heat transfer fluids and energy storage applications, respectively.

Author Contributions: Conceptualization, D.R.-G. and P.G.-R.; data curation, E.C.-A.; investigation, D.R.-G., M.d.R.R.-L., E.C.-A., D.P.D., R.B.-V. and Z.C.-H.; methodology, D.R.-G., M.d.R.R.-L., E.C.-A., and D.P.D.; project administration, R.B.-V.; supervision, P.G.-R.; writing—original draft, D.R.-G.; writing—review and editing, P.G.-R.

Funding: ICN2 acknowledges support from the Severo Ochoa Program (MINECO, grant SEV-2017-0706) and funding from the CERCA Programme/Generalitat de Catalunya. Funding from the Spanish Ministry (MAT2015-68394-R and RTI2018-099826-B-I00, MCIU/AEI/FEDER, UE) is also acknowledged.

Acknowledgments: This article is based on work from the COST Action Nanouptake, supported by the COST (European Cooperation in Science and Technology, www.cost.eu).

Conflicts of Interest: The authors declare no conflict of interest.

References

1. Novoselov, K.S.; Fal, V.I.; Colombo, L.; Gellert, P.R.; Schwab, M.G.; Kim, K. A roadmap for graphene. *Nature* **2012**, *490*, 192–200. [CrossRef] [PubMed]
2. Bonaccorso, F.; Colombo, L.; Yu, G.; Stoller, M.; Tozzini, V.; Ferrari, A.C.; Pellegrini, V. Graphene, related two-dimensional crystals, and hybrid systems for energy conversion and storage. *Science* **2015**, *347*. [CrossRef] [PubMed]
3. Sadeghinezhad, E.; Mehrali, M.; Saidur, R.; Mehrali, M.; Latibari, S.T.; Akhiani, A.R.; Metselaar, H.S.C. A comprehensive review on graphene nanofluids: Recent research, development and applications. *Energy Convers. Manag.* **2016**, *111*, 466–487. [CrossRef]
4. Taylor, R.; Coulombe, S.; Otanicar, T.; Phelan, P.; Gunawan, A.; Lv, W.; Tyagi, H. Small particles, big impacts: A review of the diverse applications of nanofluids. *J. Appl. Phys.* **2013**, *113*, 1. [CrossRef]
5. Kakaç, S.; Pramuanjaroenkij, A. Review of convective heat transfer enhancement with nanofluids. *Int. J. Heat Mass Transf.* **2009**, *52*, 3187–3196. [CrossRef]
6. Rueda-Garcia, D.; Cabán-Huertas, Z.; Sánchez-Ribot, S.; Marchante, C.; Benages, R.; Dubal, D.P.; Ayyad, O.; Gómez-Romero, P. Battery and supercapacitor materials in flow cells. Electrochemical energy storage in a LiFePO4 / reduced graphene oxide aqueous nano fluid. *Electrochim. Acta* **2018**, *281*, 594–600.
7. Sen, S.; Govindarajan, V.; Pelliccione, C.J.; Wang, J.; Miller, D.J.; Timofeeva, E.V. Surface Modification approach to TiO_2 nanofluids with high particle concentration, low viscosity, and electrochemical activity. *ACS Appl. Mater. Interfaces* **2015**, *7*, 20538–20547. [CrossRef]
8. Dubal, D.P.; Gomez-Romero, P. Electroactive graphene nanofluids for fast energy storage. *2D Mater.* **2016**, *3*, 031004. [CrossRef]
9. Dubal, D.P.; Ayyad, O.; Ruiz, V.; Gómez-Romero, P. Hybrid energy storage: The merging of battery and supercapacitor chemistries. *Chem. Soc. Rev.* **2015**, *44*, 1777–1790. [CrossRef]
10. Dubal, D.P.; Rueda-Garcia, D.; Marchante, C.; Benages, R.; Gomez-Romero, P. Hybrid graphene-polyoxometalates nanofluids as liquid electrodes for dual energy storage in novel flow cells. *Chem. Rec.* **2018**, *18*, 1076–1084. [CrossRef]
11. Rodríguez-Laguna, M.D.R.; Castro-Alvarez, A.; Sledzinska, M.; Maire, J.; Costanzo, F.; Ensing, B.; Chávez-Ángel, E. Mechanisms behind the enhancement of thermal properties of graphene nanofluids. *Nanoscale* **2018**, *10*, 15402–15409. [CrossRef] [PubMed]
12. Hermann, H.; Schubert, T.; Gruner, W.; Mattern, N. Structure and chemical reactivity of ball-milled graphite. *Nanostructured Mater.* **1997**, *8*, 215–229. [CrossRef]
13. Oh, D.W.; Jain, A.; Eaton, J.K.; Goodson, K.E.; Lee, J.S. Thermal conductivity measurement and sedimentation detection of aluminum oxide nanofluids by using the 3ω method. *Int. J. Heat Fluid Flow* **2008**, *29*, 1456–1461. [CrossRef]
14. Azmi, W.H.; Sharma, K.V.; Mamat, R.; Najafi, G.; Mohamad, M.S. The enhancement of effective thermal conductivity and effective dynamic viscosity of nanofluids—A review. *Renew. Sustain. Energy Rev.* **2016**, *53*, 1046–1058. [CrossRef]
15. Nan, C.W.; Shi, Z.; Lin, Y. A simple model for thermal conductivity of carbon nanotube-based composites. *Chem. Phys. Lett.* **2003**, *375*, 666–669. [CrossRef]
16. Mehrali, M.; Sadeghinezhad, E.; Latibari, S.T.; Kazi, S.N.; Mehrali, M.; Zubir, M.N.B.M.; Metselaar, H.S.C. Investigation of thermal conductivity and rheological properties of nanofluids containing graphene nanoplatelets. *Nanoscale Res. Lett.* **2014**, *9*, 15. [CrossRef]
17. Rodríguez-Laguna, M.D.R.; Romero, P.G.; Torres, C.M.S.; Chavez-Angel, E. Modification of the raman spectra in graphene-based nanofluids and its correlation with thermal properties. *Nanomaterials* **2019**, *9*, 804. [CrossRef]

18. Lee, J.; Weingarth, D.; Grobelsek, I.; Presser, V. Use of surfactants for continuous operation of aqueous electrochemical flow capacitors. *Energy Technol.* **2016**, *4*, 75–84. [CrossRef]
19. Dubal, D.P.; Suarez-Guevara, J.; Tonti, D.; Enciso, E.; Gomez-Romero, P. A high voltage solid state symmetric supercapacitor based on graphene-polyoxometalate hybrid electrodes with a hydroquinone doped hybrid gel-electrolyte. *J. Mater. Chem.* **2015**, *3*, 23483–23492. [CrossRef]
20. Ding, Y.; Zhang, C.; Zhang, L.; Zhou, Y.; Yu, G. Molecular engineering of organic electroactive materials for redox flow batteries. *Chem. Soc. Rev.* **2018**, *47*, 69–103. [CrossRef]
21. Chen, R. Toward High-Voltage, Energy-Dense, and durable aqueous organic redox flow batteries: Role of the supporting electrolytes. *ChemElectroChem* **2019**, *6*, 603–612. [CrossRef]
22. González, A.; Goikolea, E.; Barrena, J.A.; Mysyk, R. Review on supercapacitors: Technologies and materials. *Renew. Sustain. Energy Rev.* **2016**, *58*, 1189–1206. [CrossRef]
23. Nitta, N.; Wu, F.; Lee, J.T.; Yushin, G. Li-ion battery materials: Present and future. *Mater. Today* **2015**, *18*, 252–264. [CrossRef]
24. Kamila, S.; Mohanty, B.; Samantara, A.K.; Guha, P.; Ghosh, A.; Jena, B.; Jena, B.K. Highly active 2D layered MoS 2-rGO hybrids for energy conversion and storage applications. *Sci. Rep.* **2017**, *7*, 8378. [CrossRef] [PubMed]
25. Sen, S.; Chow, C.M.; Moazzen, E.; Segre, C.U.; Timofeeva, E.V. Electroactive nanofluids with high solid loading and low viscosity for rechargeable redox flow batteries. *J. Appl. Electrochem.* **2017**, *47*, 593–605. [CrossRef]
26. Friedl, J.; Holland-Cunz, M. V.; Cording, F.; Pfanschilling, F. L.; Wills, C.; McFarlane, W.; Stimming, U. Asymmetric polyoxometalate electrolytes for advanced redox flow batteries. *Energy Environ. Sci.* **2018**, *11*, 3010–3018. [CrossRef]
27. Joseph, A.; Xavier, M.M.; Fal, J.; Żyła, G.; Sasi, S.; Nair, P.R.; Mathew, S. Synthesis and electrochemical characterization of electroactive IoNanofluids with high dielectric constants from hydrated ferrous sulphate. *Chem. Commun.* **2019**, *55*, 83–86. [CrossRef]

Article

Prediction of Contact Angle of Nanofluids by Single-Phase Approaches †

Nur Çobanoğlu [1,2,*], Ziya Haktan Karadeniz [3], Patrice Estellé [4,*], Raul Martínez-Cuenca [5] and Matthias H. Buschmann [2]

1 Graduate School of Natural and Applied Sciences, İzmir Kâtip Çelebi University, 35620 İzmir, Turkey
2 Institut für Luft- und Kältetechnik gGmbH Dresden, 01309 Dresden, Germany; Matthias.Buschmann@ilkdresden.de
3 Department of Mechanical Engineering, İzmir Kâtip Çelebi University, 35620 İzmir, Turkey; zhaktan.karadeniz@ikcu.edu.tr
4 Univ Rennes, LGCGM, EA3913, F-35000 Rennes, France
5 Departamento de Ingeniería Mecánica y Construcción, Universitat Jaume I, 12071 Castelló de la Plana, Spain; rcuenca@uji.es
* Correspondence: ncobanoglu93.phd@gmail.com (N.Ç.); patrice.estelle@univ-rennes1.fr (P.E.)

† This article is an extended version of our paper published in 1st International Conference on Nanofluids (ICNf) and 2nd European Symposium on Nanofluids (ESNf), Castellón, Spain, 26–28 June 2019; pp. 162–166.

Received: 30 October 2019; Accepted: 27 November 2019; Published: 29 November 2019

Abstract: Wettability is the ability of the liquid to contact with the solid surface at the surrounding fluid and its degree is defined by contact angle (CA), which is calculated with balance between adhesive and cohesive forces on droplet surface. Thermophysical properties of the droplet, the forces acting on the droplet, atmosphere surrounding the droplet and the substrate surface are the main parameters affecting on CA. With nanofluids (NF), nanoparticle concentration and size and shape can modify the contact angle and thus wettability. This study investigates the validity of single-phase CA correlations for several nanofluids with different types of nanoparticles dispersed in water. Geometrical parameters of sessile droplet (height of the droplet, wetting radius and radius of curvature at the apex) are used in the tested correlations, which are based on force balance acting on the droplet surface, energy balance, spherical dome approach and empirical expression, respectively. It is shown that single-phase models can be expressed in terms of Bond number, the non-dimensional droplet volume and two geometrical similarity simplexes. It is demonstrated that they can be used successfully to predict CA of dilute nanofluids' at ambient conditions. Besides evaluation of CA, droplet shape is also well predicted for all nanofluid samples with ±5% error.

Keywords: contact angle; nanofluid; Bond number

---

1. Introduction

Wettability is the property of a solid surface contacted with a liquid within a surrounding fluid (liquid or gas) and its quantity is defined by the contact angle (CA). Wettability is important for the industrial applications such as phase change heat transfer, oil recovery and liquid coating. Contact angle (CA) is the angle between liquid–gas (or liquid–liquid) interface and liquid–solid interface for a droplet on a solid surface. The CA strongly depends on solid, liquid and surrounding fluid properties. If CA is smaller than 90°, it means that the liquid wets the solid surface. For CA greater than 90°, the liquid has lower wettability. CA is also dependent on liquid type and surface properties. Surface material including chemical composition and morphology, roughness and contamination make the CA and thus wetting properties. Since CA is predicted by the balance of interfacial/surface forces at triple point where solid, liquid and surrounding fluid contact, surrounding fluid properties play a role in wettability and CA evaluation. Besides these main parameters, temperature is another important affecting parameter for CA. Increase of temperature results in evaporation of the droplet and modifies CA.

It was also found that addition of nanoparticles (at least one dimension <100 nm) into conventional heat transfer fluids affects their thermophysical properties and results in interesting enhanced heat transfer ability. These suspensions were called as nanofluid (NF) [1]. In order to increase the markets' uptake of nanofluids, thermophysical properties of nanofluids have been studied widely [2–9]. However, surface tension and wettability behavior of NF is still at the beginning compared to other properties. Estellé et al. [10] have reviewed the surface tension and wettability and have concluded that nanoparticle nature, shape, size, content and concentration are governing parameters on surface tension and wettability. Some studies reported that the addition of nanoparticles increases the CA [11–14]. However, also a reduction in CA has also been observed in [15–17]. Lu et al. [18] also mentioned that increase in concentration decreases the spreading area of NF. However, effect of addition of nanoparticles changes with temperature [19]. Chinnam et al. [20] analyzed particle size, concentration and temperature effect on CA of NFs and developed a correlation. CA was also recently described as a relevant thermophysical property for dilute nanofluids [19]. As a result of round robin test carried out by nine independent European institutes, physically founded CA correlations dependent on temperature and droplet volume are developed from the experimental data. With zero-volume approximation, it was shown that limiting contact angle is dependent only on temperature at certain substrate and certain atmosphere.

More recently, wetting behavior of molten salt nanofluids and phase-change material nano-emulsions was studied by Grosu et al. [21]. They used wettability control to propose new methods to predict specific heat and thermal conductivity of molten salts and their nanofluids accurately. Moreover, Cabaleiro et al. [22] found that addition of a phase change material into water decreases the CA up to 52%.

Generally, measurement of CA is not easy and can have limited accuracy due to the stability of nanofluid and measurement methods. For practical applications, it is important to obtain accurate CA data of NF and models could be useful. So, the aim of this study is to evaluate the relevance of single-phase liquids models for the prediction of contact angle and droplet shape of nanofluids. Such an investigation is here reported for the first time where a comparison of different models is proposed for a large variety of nanofluids. A previous study [23] aimed to predict CA of dilute NFs by using single phase models, which were based on force balance and energy balance on the droplet surface. The previous study [23] was extended by considering a theoretical and an empirical model to improve the accuracy. Theoretical model was a spherical dome approach and empirical model was developed by Wong et al. [24]. Moreover, prediction of droplet shape was studied. All models were expressed in terms of Bond number (*Bo*) and geometrical similarity simplexes of droplets.

2. Materials and Methods

2.1. Nanofluids

To assess the relevance of different single-phase models for the prediction of nanofluid contact angle, experimental data for distilled water (DIW) as a single-phase reference fluid and base fluid of nanofluids with GO, Au, SiO_2 and Al_2O_3 nanoparticles were utilized. Au, Al_2O_3 and GO NFs here employed were fully characterized in [19] where preparation and thermophysical properties evaluation including density are also reported.

For preparation of SiO_2 NF, silica raw material with density of 2000 kg/m^3 at 20 °C (SIPERNAT® 22S, Evonik Industries AG, Germany) was dispersed unfractionated. NF had pH- value of 10.5 with stabilization by KOH. Mean agglomerate size of particles was measured as 177 nm by using Dynamic Light Scattering analysis (Zetasizer Nano ZS, Malvern Instruments GmbH, Germany). Density of SiO_2 NF was calculated with the mixture rule (Equation (1)) [25]. Thermal conductivity of SiO_2 nanofluid was measured by using 3ω method [26]. Measurements were conducted at 24 °C with three repetitions. Main information for all nanofluids are compiled in Table 1 showing the difference in concentration, size, shape and density of nanoparticles, and density and thermal conductivity of NF.

$$\rho_e = \left(1 - \phi_{p,v}\right)\rho_f + \phi_{p,v}\rho_p. \tag{1}$$

Table 1. Characteristics of working fluids.

Nanoparticle	Particle Size	Particle Shape	ϕ_p	Density of Nanoparticle (kg/m^3)	Density of NF (kg/m^3)	Thermal Conductivity Ratio (k_{NF}/k_{DIW})
Gold (Au)	Particle diameter: 8.34 nm	Spherical	0.001 wt. % (0.000052 vol.%)	1,9300	997.25	0.999
Silica (SiO_2)	Particle diameter: 117 nm	Spherical	3.935 wt. % (2 vol.%)	2000	1017.06	1.008
Graphene oxide (GO)	Extension of particle: 770 to 900 nm Thickness: 2 nm to 10 nm	Flake	0.01 wt. % (0.005679 vol.%)	1500–1900	997.25	0.9964
Alumina (Al_2O_3)	Particle Diameter: 123 ± 2 nm	Spherical	0.4 wt. % (0.1 vol. %)	3987	997.25	0.9961

2.2. Experiments on Contact Angle Measurements and Determination of Geometrical Parameters

Experiments were performed in different institutes. The CA data of DIW, Au, Al_2O_3 and GO NFs were collected at İzmir Kâtip Çelebi University (İKÇÜ), Universitat Jaume I Castelló (UJI) and Université Rennes 1 (UR1) and were taken from [19]. CA of SiO_2 nanofluid was measured at ILK-Dresden (ILK). Sessile drop method was used in all institutes. Devices used in institutes and experimental conditions such as temperature T, droplet volume V and relative humidity RH used by each institute for the different liquids are described in Table 2.

Table 2. Devices, temperature, relative humidity and droplet volume of experimental study in each institute.

Institutions & Devices	Working Fluids				
	DIW	Au NF	GO NF	SiO_2 NF	Al_2O_3 NF
İKÇÜ Attention Theta Goniometer (Biolin Scientific, (Sweden/Finland))	T = 24.2 °C RH = 40% V = 4.6 µL	T = 23.7 °C RH = 40% V = 9.8 µL	T = 23.1 °C RH = 36% V = 4.1 µL		
ILK Lab-made device	T = 22.0 °C RH = 67% V = 10 µL			T = 25.0 °C RH = 64.5 % V = 10 µL	
UJI Lab-made device	T = 24.0 °C RH = 54% V = 5.1–71.1 µL	T = 24.0 °C RH = 54% V = 5.3–68.6 µL	T = 24.0 °C RH = 54% V = 8.4–35.4 µL		T = 24.0 °C RH = 54% V = 5.5–28.8 µL
UR1 DSA-30 Drop Shape Analyzer (KRÜSS GmbH, Germany)	T = 21.0 °C RH = 24% V = 22.3 µL	T = 21.0 °C RH = 24% V = 24.1 µL	T = 21.0 °C RH = 24% V = 34.3 µL		T = 21.0 °C RH = 24% V = 20.9 µL

Main steps of the CA measurements performed by each institute are described in the following paragraph. The reader is also referred to [19] for more details. Attention Theta Goniometer (Biolin Scientific, (Sweden/Finland)) was used in the experiments carried out by İKÇÜ. After calibration of the device, surface tension of liquids was measured and then introduced to device software as property to find CA. In 10 s, 125–140 images of droplet were captured. CA was evaluated by fitting the droplet shape with the Young-Laplace equation.

UR1 calibrated the commercial device (DSA-30 Drop Shape Analyzer (KRÜSS GmbH, Germany)) from gauges with known CAs of 30°, 60° and 120°, and a maximum relative deviation was obtained as 0.52%. Droplet was positioned at the surface by using 15-gauge needle with an outer diameter of

1.835 mm. After positioning, CA values were measured in a few seconds. Further information on setup, experimental procedure and image analysis was given in [19].

Lab-made device developed by UJI was combination of camera, syringe, indirect LED light source and micrometer carrier. The droplet regulator, which holds the syringe, was controlled by a micrometer carrier. Syringe was used to position the droplet on the solid surface at a controlled volume. LED panel light was used to obtain high contrast by background illumination. This device allows the prediction of CA and volume of the droplet by image processing.

CA was measured at ILK by using the lab-made device [19]. In the setup [19], the droplet was positioned by a micropipette (VWR Pipettor 2–20 μL) on the solid substrate. Digital reflex camera (Canon EOS 40D) equipped with close-up lens (Tamron, SP 90 mm F/2.8, Di MACRO 1:1, VC USD; Japan) was used to capture the droplet images. In order to obtain accurate CA, contrast of images improved by using light disperser and light source with 129 LEDs. Images of droplet were stored at a laptop and contact angle of each droplet was predicted by using ImageJ [27] software with extension "drop analysis" [28]. Contact angle of the droplet was determined for both sides and mean contact angle was used. Experiments of CA measurement were conducted in ILK for constant volume at 10 μL.

CA of DIW and NFs was measured on the solid stainless steel substrate with thickness of 5 mm and an averaged roughness of 1.4 μm. For each institution, solid substrates were manufactured from single round stock with same lathe to provide same material and surface quality for all substrates. Properties of the solid substrate were presented in [19] with details.

In addition to CA, surface tension (σ_{lg}) of the working fluids was evaluated experimentally at UR1 (DIW, Au NF, Al_2O_3 and GO NF) and İKÇÜ (SiO_2 NF; Table 3). In both institutes, the pendant drop method was employed to define the σ_{lg}. DSA-30 Drop Shape Analyzer (KRÜSS GmbH, Germany) and Biolin Scientific, (Sweden/Finland) were used at UR1 and İKÇÜ, respectively. In İKÇÜ, surface tension of DIW was measured before SiO_2 NF. For both working fluids, the volume was kept constant at 10 μL and three measurements were carried out. There is no significant deviation between replicates and between experimental values of DIW with NIST data [29] at a given temperature. The comprehensive description of the experimental procedure used for the surface tension measurement at UR1 can be found in [30].

Table 3. Surface tensions of the working fluids.

Working Fluid	T (°C)	RH (%)	σ_{lg} (mN/m)	Standard Deviation
DIW	21	24	72.960	0.06
GO NF	21	24	73.345	0.125
Au NF	21	24	72.77	0.13
Al_2O_3 NF	21	24	72.005	0.255
SiO_2 NF	22.8	40	70.13	0.25

Finally, the geometrical parameters (δ, r_d and R_0) used in the models that are described in the next section were obtained by CA image analysis. Such an analysis has been performed with an image processing program called Fiji [31] as shown in Figure 1. Pixel-to-mm ratio with respect to a reference dimension for the images from the institutes is defined to spatial calibration of each data set.

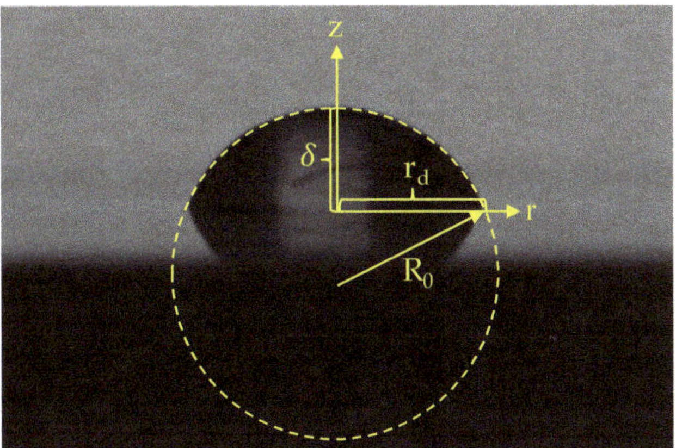

Figure 1. Geometrical parameters on droplet.

2.3. Single-Phase Models for Prediction of Contact Angle

In this section, the models developed for the CA evaluation of single-phase liquids are presented. The first model considered was developed by Vafaei and Podowski [32], which is based on the force balance of gravity and surface forces on the droplet surface. Reason behind the gravity force is to the hydrostatic force of the variable droplet height and the surface forces are result of external pressure and surface tension. Vafaei and Podowski [32] proposed a model for the prediction of droplet volume that writes as follow:

$$V = \frac{2\pi\sigma_{lg}r_d^2}{\rho g}\left(\frac{\rho g\delta}{2\sigma_{lg}} + \frac{1}{R_0} - \frac{sin\theta_d}{r_d}\right). \tag{2}$$

Here ρ denotes density and V are droplet baseline radius and volume. As geometrical parameters r_d, δ and R_0 are the droplet baseline radius, apex height and curvature, respectively. θ_d is the contact angle of the droplet. Droplet volume equation (2) was reformulated by using non-dimensional numbers: Bond number (Bo), which is the ratio of gravitational and surface forces, two geometrical similarity simplexes—G_1 and G_2—which describe the droplet geometry and V^* as the non-dimensional droplet volume.

$$\begin{aligned} sin\theta_d &= \tfrac{1}{2}Bo\,G_1 + G_2 - Bo\,V^*; \\ G_1 &= \tfrac{\delta}{r_d}; \\ G_2 &= \tfrac{r_d}{R_0}; \\ V^* &= \tfrac{V}{2\pi r_d{}^3}; \end{aligned} \tag{3}$$

where Bo is

$$Bo = \rho g r_d{}^2/\sigma_{lg}. \tag{4}$$

In the definition of Bo number characteristic length equals to r_d, which is defined by Kuiken [33]. Moreover, Stacy [34] reported that characteristic length equals to radius at the droplet equator and wetting radius for non-wetting droplets and for wetting droplets, respectively.

The second model we have considered was proposed by Yonemoto and Kunugi [35] (YK model), which is based on the energy balance for various-sized droplets on a solid surface. This correlation writes

$$\frac{\rho g\delta V}{2} = \pi r_d{}^2\sigma_{lg}(1 - cos\theta_s) - \pi r_d\delta\sigma_{lg}sin\theta_s. \tag{5}$$

A non-dimensional form is obtained by dividing (5) by the gravitational potential $\rho g\delta V/2$.

$$\begin{aligned} 1 &= X^* - Y^*; \\ X^* &= \frac{2\pi r_d^2}{\delta V}\frac{\sigma_{lg}(1-cos\theta_s)}{\rho g} = \frac{(1-cos\theta_s)}{Bo\ V^*\ G_1}; \\ Y^* &= \frac{2\pi r_d}{V}\frac{\sigma_{lg}sin\theta_s}{\rho g} = \frac{sin\theta_s}{Bo\ V^*}. \end{aligned} \tag{6}$$

YK model is used for only spherical droplets. Therefore, employing the correlation for $\theta_s = \arcsin\left(\frac{r_d \sin\theta_d}{r_{ds}}\right)$ given in [32] Equation (5) is rewritten. The contact angle for spherical droplets is denoted by θ_s and its baseline radius by r_{ds}.

$$1 + cos\theta_s = \frac{sin^2\theta_d}{G_1(BoV^* + sin\theta_d)}. \tag{7}$$

Moreover, the spherical dome model (SD model), which is based on the tangent of only dimensions of the droplet (δ and r_d), analyzed in [24] was rewritten in terms of geometrical similarity simplex, G_1, and validity was studied for nanofluids.

$$\theta_d = 2arctan\left(\frac{\delta}{r_d}\right) = 2arctan(G_1). \tag{8}$$

Finally, Wong et al. [24] (W model) developed an empirical formulation, expressed in Equation (9), to predict contact angle of liquids, which was reported as being valid for any substrate and droplet materials. This empirical model uses the contact angle, droplet volume, height and width ($W = 2r_d$) and allows us to find two of them by using known other two parameters.

$$\begin{aligned} \theta_d = {} & 1.41 - 0.678\delta + 0.143W + 0.307\delta^2 - 0.0757\delta W - 0.00457W^2 \\ & -0.0453\delta^2 W + 0.0310W^2\delta - 0.00457W^3. \end{aligned} \tag{9}$$

2.4. Prediction of Droplet Shape

In addition to CA prediction from single phase models, the sessile droplet shape of NFs is compared to the following approach. Thus, Vafaei and Podowski [32] reported that droplet shape could be predicted by force balance on the droplet surface as well as CA as following:

$$z = \delta - \frac{\sigma_{lg}}{\rho g} r^2 \left(\frac{3c_1}{r} + 4c_2 + 5c_3 r + 6c_4 r^2\right). \tag{10}$$

Coefficients (c_1, c_2, c_3, c_4) were determined by applying boundary conditions:

$$z(r_d) = 0. \tag{11}$$

$$\left.\frac{dz}{dr}\right|_{r=0} = 0. \tag{12}$$

$$\left.\frac{d^2z}{dr^2}\right|_{r=0} = -\frac{1}{R_0}. \tag{13}$$

$$V = \int_0^\delta \pi r^2 dz. \tag{14}$$

Applying one more boundary condition $z(-r_d) = 0$ and power series assumption ($c_3 = 0$), Equation (10) was re-arranged by using geometrical similarity simplexes.

$$\begin{aligned} \frac{z}{r_d} &= G_1 - \frac{G_3G_4}{2} - G_1G_4{}^4 + \frac{G_3G_4{}^3}{2}; \\ G_3 &= \frac{r}{R_0}; \\ G_4 &= \frac{r}{r_d}. \end{aligned} \tag{15}$$

Considering the origin is at the center of the wetting circle, r is the horizontal axis and z is the vertical axes of the droplet (Figure 1).

3. Results and Discussion

In this study, validity of contact angle models, which were developed for single-phase fluids, was investigated. In the figures, filled symbols (•) were for models and empty symbols (○) were used for experimental results. Due to the measurement uncertainties and in order to evaluate their influence, numerical results from the models were corrected with increasing R_0 by 2% and it was presented as a red cross in the figures (×).

Error between experimental and predicted CA was determined by using mean absolute percentage error (MAPE) and was denoted as Ω in the study. It is found as:

$$\Omega = \frac{100}{n}\sum_{t=1}^{n}\left|\frac{\theta_{dexp,t} - \theta_{dpred,t}}{\theta_{dexp,t}}\right|. \quad (16)$$

3.1. Contact Angle Prediction with Single-Phase Models

Figures 2 and 3 show the validity of VP model for DIW and the different nanofluids. DIW as expected for a single-phase liquid has good agreement with the model (for various volume Ω = 2.32% and for constant volume Ω = 6.34%). Due to the low concentration and the small particle size, the Au and Al_2O_3 NF shows also reasonable good agreement with Ω = 2.07% and Ω = 4.07%, respectively. However, the VP model seems to be not valid for GO (Ω = 8.52%) and SiO_2 NFs (Ω = 30.46%). Large graphene flakes in GO NF and higher concentration (3.935 wt.%) of SiO_2 NF may be the reason of preventing the application of the VP model.

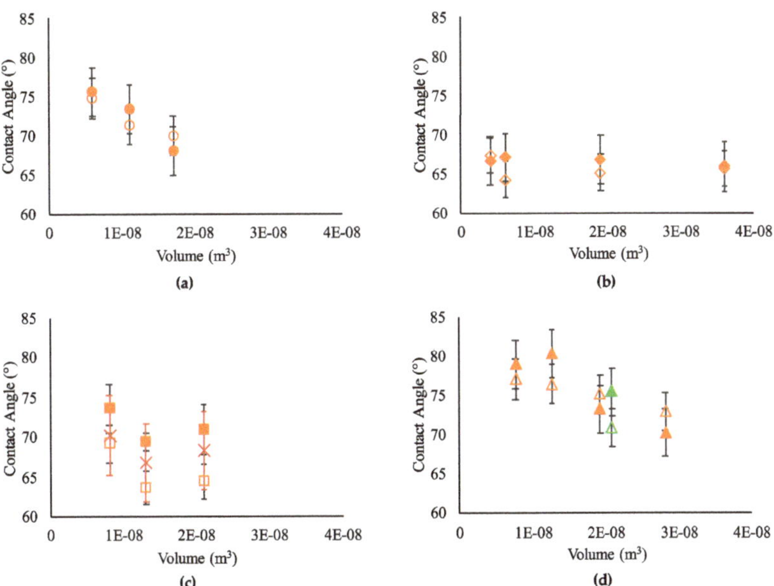

Figure 2. VP model validity for fluids at various volumes. Working fluids are (**a**) distilled water (DIW), (**b**) Au nanofluid (NF), (**c**) GO NF and (**d**) Al_2O_3 NF. Colors indicate: orange—Universitat Jaume I Castelló (UJI) and green—Université Rennes 1 (UR1).

It was found, that the VP model was very sensitive against tiny experimental errors following the determination of R_0. To illustrate this, R_0 was increased artificially by just 2% and additional results presented in Figure 2 for GO NF, and Figure 3 for DIW and SiO_2. CA data of GO NF fit much better with the VP model with this correction. Ω decreased from 8.52% to 4.19% for GO NF. Unlike GO NF, CA of SiO_2 NF was not affected from the increase of R_0 (Ω = 27.79%). This shows the influence of nanoparticle content on the applicability of V–P model with nanofluids.

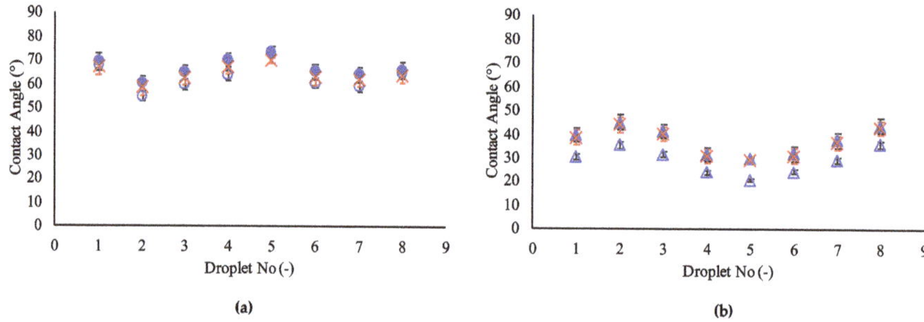

Figure 3. VP model validity for fluids at constant volume. Working fluids are (**a**) DIW and (**b**) SiO_2 NFs.

The modified YK model was in good agreement with experimental results for DIW, Au, Al_2O_3 and GO nanofluids. As shown in Figure 4, all the data was in the range of the ±10% error band. However, the results for the SiO_2 NF (Figure 5) show again that a single-phase model was not appropriate for a highly concentrated nanofluid.

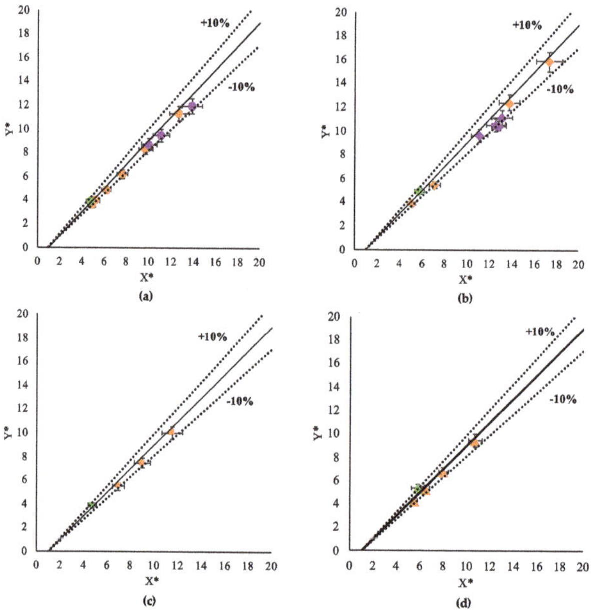

Figure 4. Yonemoto and Kunugi (YK) model validity for fluids at various volumes. Working fluids are (**a**) DIW, (**b**) Au NF, (**c**) GO NF and (**d**) Al_2O_3 NF. Colors indicate: orange—UJI, green—UR1 and purple—İzmir Kâtip Çelebi University (İKÇÜ).

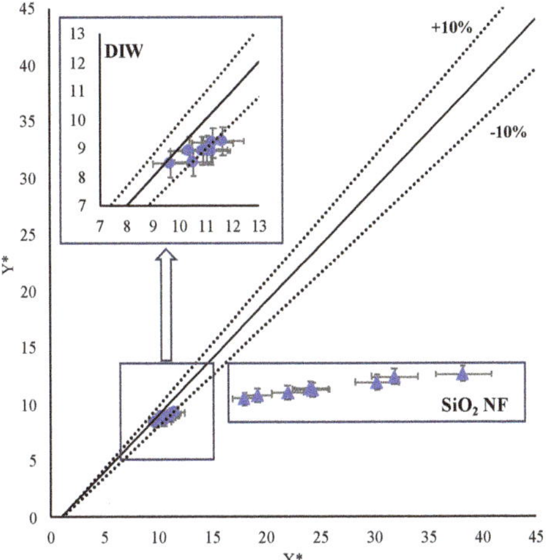

Figure 5. YK model validity for fluids at constant volume.

Figures 6 and 7 show that the SD model was valid for DIW (Ω = 4.71%), Al_2O_3 NF (Ω = 3.78%) and Au NF (Ω = 7.18%). Vafaei and Podowski [32] mentioned that as volume decreases the shape of the droplet is more spherical. Although the volume range was narrow for this study, change of the droplet shape with volume was obviously clear. For GO nanofluid at the highest volume, CA was underpredicted compared to experimental results with Ω = 7.35%. The reason could be the shape and size of the graphene flakes. Different from the other theoretical models, predicted SiO_2 data with the SD model was much collapsed with the experimental data with Ω = 10.71% where DIW had Ω = 3.90%.

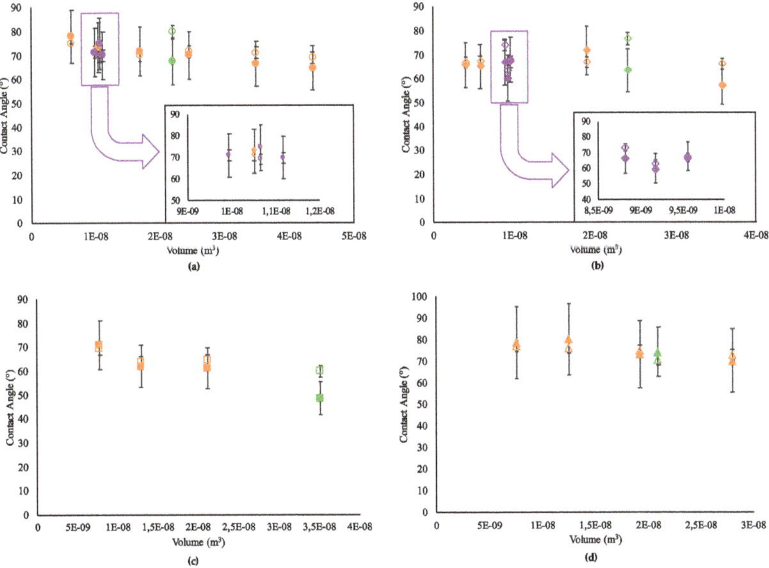

Figure 6. Spherical dome (SD) model validity for fluids at various volumes. Working fluids are (**a**) DIW, (**b**) Au NF, (**c**) GO NF and (**d**) Al_2O_3 NF. Colors indicate: orange—UJI, green—UR1 and purple—İKÇÜ.

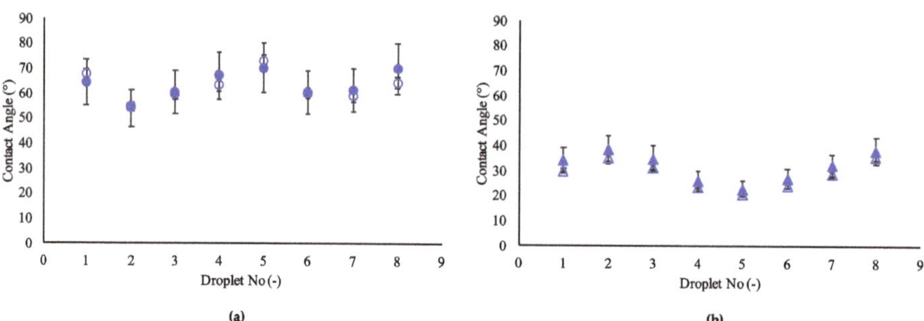

Figure 7. SD model validity for fluids at constant volume. Working fluids are (**a**) DIW and (**b**) SiO_2 NF.

In addition to these theoretical models, the empirical model developed by Wong et al. [24] allowed us to predict contact angle of nanofluids at lower concentrations (Figure 8). Ω value was calculated for DIW for various volumes, DIW at constant volume, Au NF and GO NF were 5.77%, 3.53%, 4.70% and 5.26%, respectively. CA data of Al_2O_3 NF (Ω = 9.78%) predicted by the W model is different for smaller volumes. However, increase in volume results in more accurate results with the W model. The reason could be that the concentration highly affects on CA at lower volumes for this model. Similar behavior was observed for also SiO_2 NF. The W model was not suitable for highly concentrated nanofluids as shown in SiO_2 NF (Ω = 144.46%) results in Figure 9.

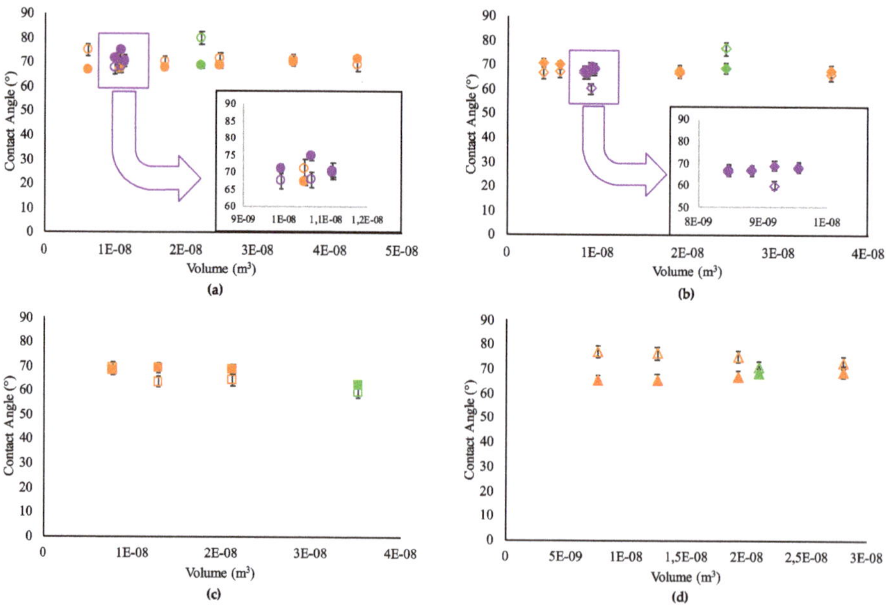

Figure 8. Wong et al. (W) model validity for fluids at various volumes. Working fluids are (**a**) DIW, (**b**) Au NF, (**c**) GO NF and (**d**) Al_2O_3 NF. Colors indicate: orange—UJI, green—UR1 and purple—İKÇÜ.

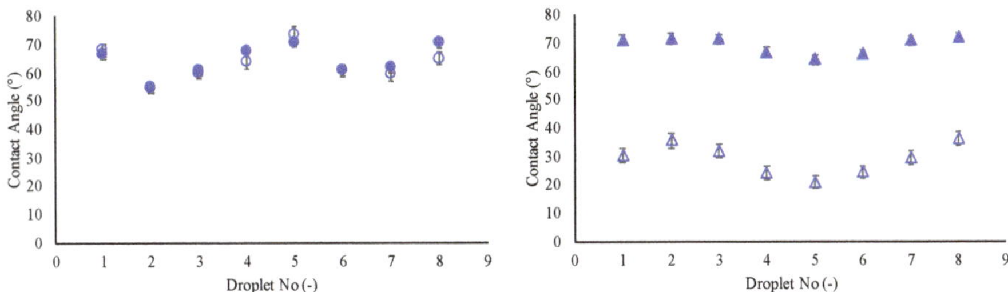

Figure 9. W model validity for fluids at constant volume. Working fluids are (**a**) DIW and (**b**) SiO_2 NF.

3.2. *Droplet Shape Prediction*

Vafaei and Podowski [32] also developed a model to predict droplet shape based on force balance (Equation (15)). This subsection presents the validity of this model for nanofluids considered in this study. For volume dependent analysis, droplets had a maximum 21 μL of volume (Figure 10). For constant volume analysis of DIW and SiO_2 NF, droplets that had maximum, minimum and average contact angles were chosen (Figure 11).

Droplet shape was predicted with ±5% error for all cases without influence of shape and concentration of nanoparticles, and volume of the droplet. This ±5% error band could include measurement errors in geometrical parameters and experimental errors.

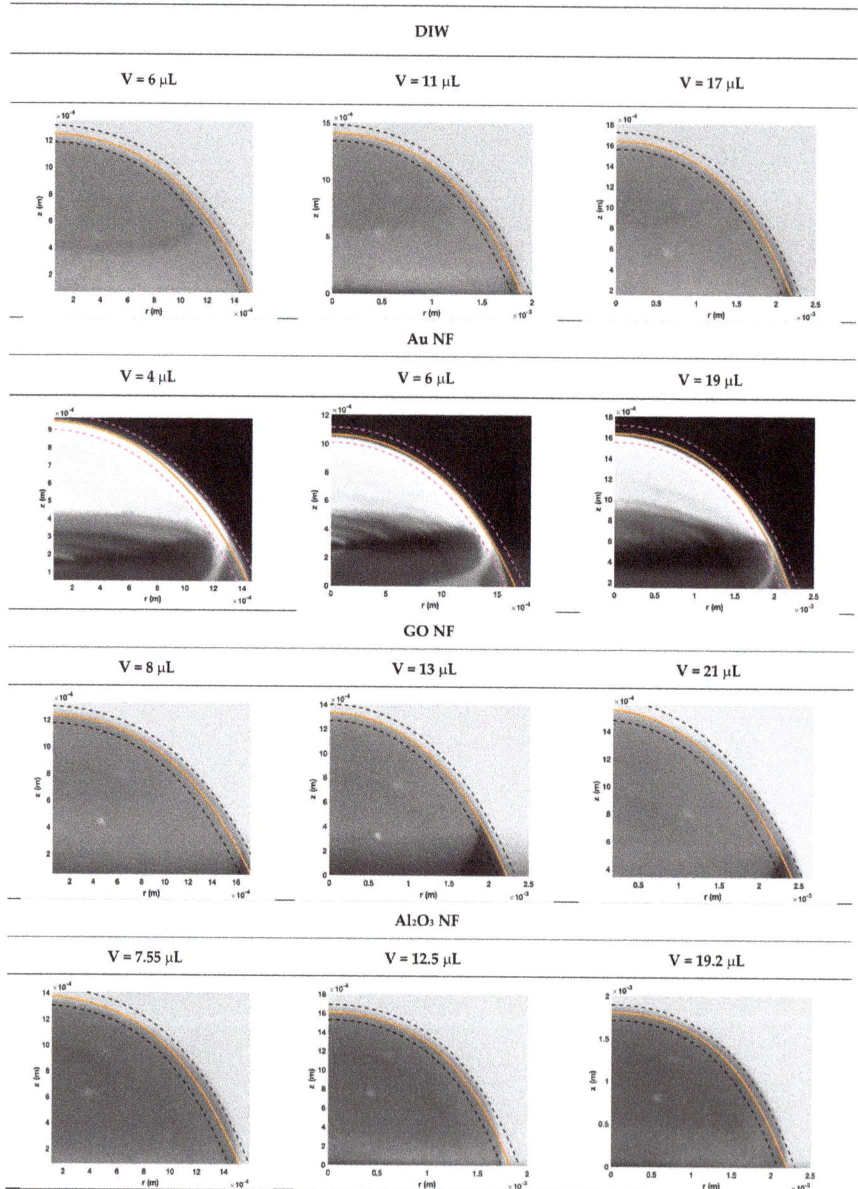

Figure 10. Droplet shape prediction at various volumes (—shows Equation (14) and—shows the ±5% errors).

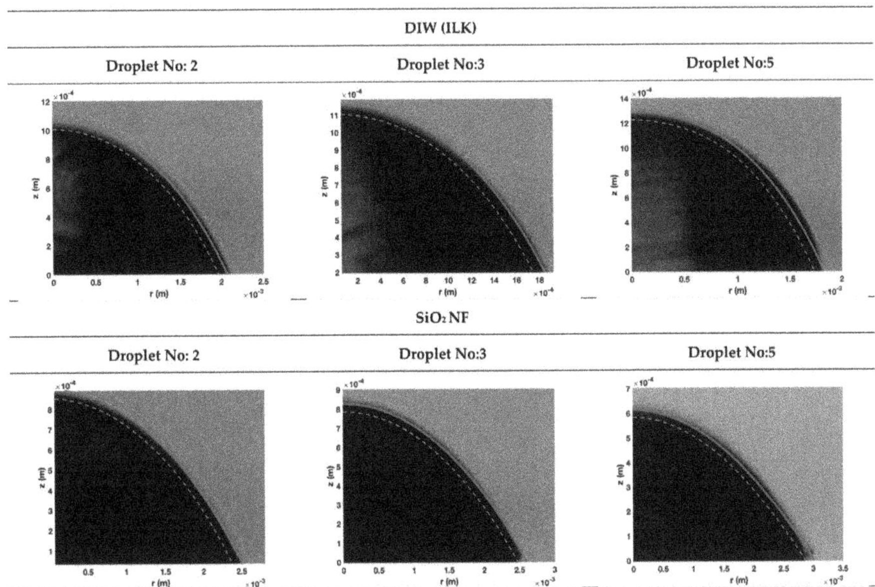

Figure 11. Droplet shape prediction at constant volume (—shows Equation (14) and—shows the ±5% errors).

4. Conclusions

The aim of this study was to evaluate the reliability of three theoretical models and one empirical model developed for single-phase liquids for describing the correlation between contact angle, geometrical parameters and forces/energy balance of nanofluid sessile droplets. Four types of water-based nanofluids were considered varying the size, shape and nature of nanoparticles as well as the concentration.

Single-phase models were expressed in terms of *Bo*, which depends on surface tension and density, non-dimensional volume V^* and geometrical simplexes and they were used for prediction of dilute NFs' CA. Although the measurement of the required geometrical parameters for prediction of CA demands high precision to obtain accurate results, it is concluded that concentration is the main factor to affect applicability of models for NFs. Although the force balance-based VP model was too sensitive to measurement errors for GO NF and CA of dilute NFs (Au, GO and Al_2O_3 NF) could be predicted by both the VP model and energy balance-based YK model. Moreover, the empirical W model was valid for dilute NFs (Au and GO NF) and higher volumes of Al_2O_3 NF. However, significant differences were obtained for highly concentrated SiO_2 NF in these theoretical models and empirical model. Additional effects like disjoining pressure, convective flows inside the droplet, etc. could be the reasons of differences in results of different models. Beside them, the SD model is suitable for almost all samples due to the spherical shape of droplets as a result of smaller volumes. However, higher error band of SD model is the limitation for accurate prediction of CA. It was also shown that the droplet shape of all NFs could be well predicted from a model based on force balance.

To conclude, further research is now needed to resolve and improve the limitation of the measurement of dimensions and for the development of a useful and generic CA model by investigation of additional possible effects such as temperature, a wider range of nanoparticle size and shape varying the nanoparticle concentration.

Author Contributions: Conceptualization, N.Ç. and M.H.B.; Methodology, N.Ç., Z.H.K., P.E. and R.M.-C.; Validation, N.Ç., P.E. and M.H.B.; Formal Analysis, N.Ç. and M.H.B.; Investigation, N.Ç., P.E. and R.M.-C.;

Resources, N.Ç., P.E. and R.M.-C.; Data Curation, N.Ç. and M.H.B.; Writing—Original Draft Preparation, N.Ç.; Writing—Review and Editing, Z.H.K., P.E., R.M.-C. and M.H.B.; Visualization, N.Ç., Z.H.K., P.E., R.M.-C. and M.H.B.; Supervision, M.H.B.; Project Administration, N.Ç., Z.H.K. and M.H.B.

Funding: N.Ç. acknowledges the EU COST Action CA15119: Overcoming Barriers to Nanofluids Market Uptake for financial support in the participation of the 1st International Conference on Nanofluids (ICNf) and the 2nd European Symposium on Nanofluids (ESNf) held at the University of Castellón, Spain during 26–28 June 2019. P.E. acknowledges the European Union through the European Regional Development Fund (ERDF), the Ministry of Higher Education and Research, the French region of Brittany and Rennes Métropole for the financial support related to the device used in this study for surface tension and contact angle measurements.

Acknowledgments: This investigation is a contribution to the COST (European Cooperation in Science and Technology) Action CA15119: Overcoming Barriers to Nanofluids Market Uptake (NanoUptake).

Conflicts of Interest: The authors declare no conflict of interest.

Nomenclature

Bo	Bond Number (–)
c	Coefficients
g	Gravitational Acceleration (m/s^2)
G	Geometrical Similarity Simplex (–)
k	Thermal Conductivity (W/mK)
r	Droplet Wetting Radius (m)
R	Radius of Curvature (m)
RH	Relative Humidity (%)
V	Volume (m^3)
W	Width of the droplet (m)
T	Temperature (°C)

Abbreviations

CA	Contact Angle
DIW	Distilled Water
İKÇÜ	Izmir Katip Çelebi University
ILK	ILK-Dresden
NF	Nanofluid
UJI	Universitat Jaume I Castelló
UR1	Université Rennes 1

Subscripts

0	At the apex
d	Droplet
e	Effective
f	Base Fluid
lg	Liquid–Gas
p	Particle
s	Spherical
v	Volumetric

Greek Letters

σ	Surface Tension (mN/m)
δ	Location of Apex (m)
ρ	Density (kg/m^3)
θ	Contact Angle (°)
Ω	Mean Absolute Percentage Error (%)
ϕ	Concentration of NF (%)

References

1. Choi, S.U.; Eastman, J.A. *Enhancing Thermal Conductivity of Fluids with Nanoparticles*; Argonne National Lab.: DuPage County, IL, USA, 1995.
2. Cabaleiro, D.; Gracia-Fernández, C.; Legido, J.; Lugo, L. Specific heat of metal oxide nanofluids at high concentrations for heat transfer. *Int. J. Heat Mass Transf.* **2015**, *88*, 872–879. [CrossRef]
3. Riazi, H.; Murphy, T.; Webber, G.B.; Atkin, R.; Tehrani, S.S.M.; Taylor, R.A. Specific heat control of nanofluids: A critical review. *Int. J. Therm. Sci.* **2016**, *107*, 25–38. [CrossRef]
4. Koca, H.D.; Doganay, S.; Turgut, A.; Tavman, I.H.; Saidur, R.; Mahbubul, I.M. Effect of particle size on the viscosity of nanofluids: A review. *Renew. Sustain. Energy Rev.* **2018**, *82*, 1664–1674. [CrossRef]
5. Murshed, S.S.; Estellé, P. A state of the art review on viscosity of nanofluids. *Renew. Sustain. Energy Rev.* **2017**, *76*, 1134–1152. [CrossRef]
6. Antoniadis, K.D.; Tertsinidou, G.J.; Assael, M.J.; Wakeham, W.A. Necessary conditions for accurate, transient hot-wire measurements of the apparent thermal conductivity of nanofluids are seldom satisfied. *Int. J. Thermophys.* **2016**, *37*, 78. [CrossRef]
7. Ahmadi, M.H.; Mirlohi, A.; Nazari, M.A.; Ghasempour, R. A review of thermal conductivity of various nanofluids. *J. Mol. Liquids* **2018**, *265*, 181–188. [CrossRef]
8. Tongkratoke, A.; Pramuanjaroenkij, A.; Kakaç, S. Numerical Study of Mixing Thermal Conductivity Models for Nanofluid Heat Transfer Enhancement. *J. Eng. Phys. Thermophys.* **2018**, *91*, 104–114.
9. Murshed, S.S.; Estellé, P. Rheological Characteristics of Nanofluids for Advance Heat Transfer. In *Advances in New Heat Transfer Fluids*; CRC Press: Boca Raton, FL, USA, 2017; pp. 227–266.
10. Estellé, P.; Cabaleiro, D.; Żyła, G.; Lugo, L.; Murshed, S.S. Current trends in surface tension and wetting behavior of nanofluids. *Renew. Sustain. Energy Rev.* **2018**, *94*, 931–944. [CrossRef]
11. Radiom, M.; Yang, C.; Chan, W.K. *Characterization of Surface Tension and Contact Angle of Nanofluids*; SPIE: Bellingham, WA, USA, 2010; Volume 7522, p. 75221D.
12. Radiom, M.; Yang, C.; Chan, W.K. Dynamic contact angle of water-based titanium oxide nanofluid. *Nanoscale Res. Lett.* **2013**, *8*, 282. [CrossRef]
13. Cieśliński, J.T.; Krygier, K.A. Sessile droplet contact angle of water–Al2O3, water–TiO2 and water–Cu nanofluids. *Exp. Therm. Fluid Sci.* **2014**, *59*, 258–263. [CrossRef]
14. Murshed, S.S.; Tan, S.-H.; Nguyen, N.-T. Temperature dependence of interfacial properties and viscosity of nanofluids for droplet-based microfluidics. *J. Phys. Appl. Phys.* **2008**, *41*, 085502. [CrossRef]
15. Harikrishnan, A.; Dhar, P.; Agnihotri, P.K.; Gedupudi, S.; Das, S.K. Wettability of complex fluids and surfactant capped nanoparticle-induced quasi-universal wetting behavior. *J. Phys. Chem. B* **2017**, *121*, 6081–6095. [CrossRef] [PubMed]
16. Jeong, Y.H.; Chang, W.J.; Chang, S.H. Wettability of heated surfaces under pool boiling using surfactant solutions and nano-fluids. *Int. J. Heat Mass Transf.* **2008**, *51*, 3025–3031. [CrossRef]
17. Prajitno, D.; Trisnawan, V.; Syarif, D. *Effect of Spreading Time on Contact Angle of Nanofluid on the Surface of Stainless Steel AISI 316 and Zircalloy 4*; IOP Publishing: Bristol, UK, 2017; Volume 196, p. 012028.
18. Lu, G.; Duan, Y.-Y.; Wang, X.-D. Experimental study on the dynamic wetting of dilute nanofluids. *Colloids Surf. Physicochem. Eng. Asp.* **2015**, *486*, 6–13. [CrossRef]
19. Hernaiz, M.; Alonso, V.; Estellé, P.; Wu, Z.; Sundén, B.; Doretti, L.; Mancin, S.; Çobanoğlu, N.; Karadeniz, Z.; Garmendia, N.; et al. The contact angle of nanofluids as thermophysical property. *J. Colloid Interface Sci.* **2019**, *547*, 393–406. [CrossRef]
20. Chinnam, J.; Das, D.; Vajjha, R.; Satti, J. Measurements of the contact angle of nanofluids and development of a new correlation. *Int. Commun. Heat Mass Transf.* **2015**, *62*, 1–12. [CrossRef]
21. Grosu, Y.; González-Fernández, L.; Nithiyanantham, U.; Faik, A. Wettability control for correct thermophysical properties determination of molten salts and their nanofluids. *Energies* **2019**, *12*, 3765. [CrossRef]
22. Cabaleiro, D.; Hamze, S.; Agresti, F.; Estellé, P.; Barison, S.; Fedele, L.; Bobbo, S. Dynamic Viscosity, Surface Tension and Wetting Behavior Studies of Paraffin–in–Water Nano–Emulsions. *Energies* **2019**, *12*, 3334. [CrossRef]
23. Çobanoğlu, N.; Karadeniz, Z.; Estellé, P.; Martínez-Cuenca, R.; Buschmann, M. On the contact angle of nanofluids. In Proceedings of the 1st International Conference on Nanofluids (ICNf2019) & 2nd European Symposium on Nanofluids (ESNf2019), Castelló, Spain, 26–28 June 2019.

24. Wong, T.I.; Wang, H.; Wang, F.; Sin, S.L.; Quan, C.G.; Wang, S.J.; Zhou, X. Empirical Formulae in Correlating Droplet Shape and Contact Angle. *Aust. J. Chem.* **2016**, *69*, 431–439. [CrossRef]
25. Murshed, S.S. Determination of effective specific heat of nanofluids. *J. Exp. Nanosci.* **2011**, *6*, 539–546. [CrossRef]
26. Turgut, A.; Sauter, C.; Chirtoc, M.; Henry, J.; Tavman, S.; Tavman, I.; Pelzl, J. AC hot wire measurement of thermophysical properties of nanofluids with 3ω method. *Eur. Phys. J. Spec. Top.* **2008**, *153*, 349–352. [CrossRef]
27. Abràmoff, M.D.; Magalhães, P.J.; Ram, S.J. Image processing with ImageJ. *Biophoton. Int.* **2004**, *11*, 36–42.
28. Stalder, A.F.; Kulik, G.; Sage, D.; Barbieri, L.; Hoffmann, P. A snake-based approach to accurate determination of both contact points and contact angles. *Colloids Surf. Physicochem. Eng. Asp.* **2006**, *286*, 92–103. [CrossRef]
29. Linstrom, P.J.; Mallard, W.G. The NIST Chemistry WebBook: A chemical data resource on the internet. *J. Chem. Eng. Data* **2001**, *46*, 1059–1063. [CrossRef]
30. Gomez-Villarejo, R.; Aguilar, T.; Hamze, S.; Estellé, P.; Navas, J. Experimental analysis of water-based nanofluids using boron nitride nanotubes with improved thermal properties. *J. Mol. Liquids* **2019**, *277*, 93–103. [CrossRef]
31. Schindelin, J.; Arganda-Carreras, I.; Frise, E.; Kaynig, V.; Longair, M.; Pietzsch, T.; Preibisch, S.; Rueden, C.; Saalfeld, S.; Schmid, B. Fiji: An open-source platform for biological-image analysis. *Nat. Methods* **2012**, *9*, 676. [CrossRef]
32. Vafaei, S.; Podowski, M. Analysis of the relationship between liquid droplet size and contact angle. *Adv. Colloid Interface Sci.* **2005**, *113*, 133–146. [CrossRef]
33. Kuiken, H. A single-parameter method for the determination of surface tension and contact angle. *Colloids Surf.* **1991**, *59*, 129–148. [CrossRef]
34. Stacy, R. Contact Angle Measurement Technique for Rough Surfaces. Ph.D. Thesis, Michigan Technological University, Houghton, MI, USA, 2009.
35. Yonemoto, Y.; Kunugi, T. Wettability model for various-sized droplets on solid surfaces. *Phys. Fluids* **2014**, *26*, 082110. [CrossRef]

Article

Stability and Thermal Properties Study of Metal Chalcogenide-Based Nanofluids for Concentrating Solar Power †

Paloma Martínez-Merino, Rodrigo Alcántara, Teresa Aguilar, Juan Jesús Gallardo, Iván Carrillo-Berdugo, Roberto Gómez-Villarejo, Mabel Rodríguez-Fernández and Javier Navas *

Departamento de Química Física, Facultad de Ciencias, Universidad de Cádiz, E-11510 Cádiz, Spain; paloma.martinez@uca.es (P.M.-M.); rodrigo.alcantara@uca.es (R.A.); mariateresa.aguilar@uca.es (T.A.); jj.gallardo@uca.es (J.J.G.); ivan.carrillo@uca.es (I.C.-B.); roberto.gomezvi@uca.es (R.G.-V.); mariaisabel.rodriferna@alum.uca.es (M.R.-F.)

* Correspondence: javier.navas@uca.es

† This paper is an extended version of our paper entitled "Stability and thermal properties study of 2D-metal chalcogenides-based nanofluids for concentrating solar power" published in 1st International Conference on Nanofluids (ICNf2019) and 2nd European Symposium on Nanofluids (ESNf2019), Castelló, Spain, 26–28 June 2019; pp. 182–185.

Received: 30 October 2019; Accepted: 2 December 2019; Published: 6 December 2019

Abstract: Nanofluids are colloidal suspensions of nanomaterials in a fluid which exhibit enhanced thermophysical properties compared to conventional fluids. The addition of nanomaterials to a fluid can increase the thermal conductivity, isobaric-specific heat, diffusivity, and the convective heat transfer coefficient of the original fluid. For this reason, nanofluids have been studied over the last decades in many fields such as biomedicine, industrial cooling, nuclear reactors, and also in solar thermal applications. In this paper, we report the preparation and characterization of nanofluids based on one-dimensional MoS_2 and WS_2 nanosheets to improve the thermal properties of the heat transfer fluid currently used in concentrating solar plants (CSP). A comparative study of both types of nanofluids was performed for explaining the influence of nanostructure morphologies on nanofluid stability and thermal properties. The nanofluids prepared in this work present a high stability over time and thermal conductivity enhancements of up to 46% for MoS_2-based nanofluid and up to 35% for WS_2-based nanofluid. These results led to an increase in the efficiency of the solar collectors of 21.3% and 16.8% when the nanofluids based on MoS_2 nanowires or WS_2 nanosheets were used instead of the typical thermal oil.

Keywords: nanofluids; heat transfer fluid (HTF); concentrating solar power (CSP); parabolic trough collector (PTC); nanowires; nanosheets; stability; thermophysical properties

1. Introduction

Over the last decades, the interest around renewable energies has increased due to the increasing energy demand and the environmental problems derived from fossil fuels combustion. In this scenario, solar energy presents a high potential to supply the primary energy demand, although the technology to harvest and store solar radiation needs to advance to make it affordable in comparison with fossil-based electricity. Currently, there are two technologies for power generation based on solar energy: concentrating solar power (CSP) and photovoltaics (PVs). The CSP systems present a higher potential to store energy in comparison with PV systems [1]. As a result, CSP with a thermal energy storage (TES) system is considered the first choice to provide electricity on a large scale, even at nighttime or on cloudy days [2,3]. The principle behind CSP consists of the reflection of solar irradiation in a small area where heat is collected by a fluid and, subsequently, used in a Rankine cycle to generate steam. Finally, the steam drives a turbine which powers a generator to produce electricity [4–6]. Among the different solar collectors, parabolic trough collectors (PTCs) are the most installed technology worldwide [7]. In PTC technology, a group of curved reflectors focus sunrays onto an absorber tube which is located in the focal line of collectors. This tube contains the heat transfer fluid (HTF) which collects and transports the thermal energy to electricity generation systems or to storage facilities [8–10].

One of the research lines to increase the efficiency in CSP plants based on PTCs is replacing the HTF (water or oil) with nanofluids. Typically, nanofluids, which are nanocolloidal suspensions of particles in a fluid, present improved thermophysical properties compared to conventional fluids. Tyagi et al. [11] predicted a 10% increment in the efficiency of solar collectors when nanofluids of aluminum are used as the working fluid. Also, an analysis of the applicability of nanofluids of graphite, aluminum, copper, and silver at tower solar power collectors has been reported with interesting results. Nanofluids based on these nanoparticles were analyzed for use in direct absorption solar collectors, and an efficiency improvement of up to 10% was found [12]. As regards PTCs, Mwesigye et al. [13] discovered an improvement in thermal efficiency of 7% with the use of a nanofluid based on Al_2O_3 nanoparticles. Nevertheless, the number of works on nanofluids prepared with the eutectic mixture of biphenyl and diphenyl oxide, which is the typical organic oil used in CSP plants, is very small. Furthermore, there are few works on nanofluids based on one-dimensional and two-dimensional nanomaterials such as those offered in this study.

In this paper, we used the eutectic mixture of biphenyl and diphenyl oxide as a base fluid to prepare nanofluids based on metal chalcogenides. Because an important challenge is to obtain stable nanofluids over time, the initial aim was to prepare bidimensional nanostructures of metal chalcogenides suspended in HTF. The highest aspect ratio of nanosheets avoids the common sedimentation observed in spherical nanoparticles. Both MoS_2 and WS_2 materials present high thermal conductivities (34.5 $Wm^{-1}K^{-1}$ and 32 $Wm^{-1}K^{-1}$) and a hexagonal structure which facilitates its exfoliation in nanosheets [14–16]. Moreover, polyethylene glycol (PEG) was used as surfactant since its efficiency in HTF systems was proven in a previous work [17]. However, unexpectedly, in the case of MoS_2 instead of nanosheets, the resulting nanomaterial obtained after liquid phase exfoliation (LPE) were nanowires. Therefore, during the present work, a comprehensive study was performed on the influence of the morphology of metal chalcogenide nanostructures in the stability and thermal improvements of nanofluids.

2. Materials and Methods

2.1. Reagents

WS_2 (nanopowder, 90 nm), MoS_2 (powder, <2 μm), and polyethylene glycol-200 (PEG) were purchased from Sigma–Aldrich. The HTF employed to prepare the nanofluids was the eutectic mixture of biphenyl (26.5%) and diphenyl oxide (73.5%) typically used in PTC plants. The commercial brand name of this eutectic mixture is Dowtherm A, and it was supplied by Dow Chemical Company©. All chemicals were used without further purification.

2.2. Preparation of MoS_2- and WS_2-Based Nanofluids

Liquid phase exfoliation was used to prepare the metal chalcogenide-based nanofluids. In this process, 15 mg of metal chalcogenide (MoS_2 or WS_2) were weighed and add to four vials. Then, 5 mL of a solution with a mass concentration of 0.20 wt.% PEG in HTF was added to each vial. The vials were introduced to an Elmasonic-P ultrasonic bath, and 80 kHz of frequency was applied for 4 h. After that, black-colored solutions were centrifuged at 1000 rpm for 10 min. The precipitates were discarded, and the supernatants were subjected to another centrifugation at 400 rpm for 10 min to remove non-exfoliated nanomaterial. The supernatant liquid obtained in the second centrifugation was the final nanofluid.

2.3. Characterization of Nanofluids

The nanofluids prepared were characterized to study their stability and thermal properties. Both types of characterization are essential to elucidate their application in CSP plants. Also, transmission electron microscopy was used to study the size and morphology of the nanostructures present in both nanofluids using a JEM-2100F microscope supplied by Jeol®. The stability of nanofluids was studied over a week by means of UV-Vis spectroscopy and dynamic light scattering (DLS). The UV-Vis spectra were registered between 400 and 850 nm using a USB2000+ spectrometer and an Ocean Optics® DH-2000-BAL halogen lamp. The evolution of the extinction coefficient was performed at 678 nm for MoS_2 and 629 nm for WS_2 where there are characteristics bands for these metal chalcogenides [18,19]. Dynamic light scattering was performed to determine the hydrodynamic particle size evolution over time. The equipment used to perform the DLS measurements was a Zetasizer Nano ZS system supplied by Malvern Instruments Ltd®.

Thermal conductivity (k) was determined using the equation $k(T) = D(T)\cdot C_P(T)\cdot \rho(T)$, where D is the thermal diffusivity, C_P is the isobaric specific heat, and ρ is the density which was defined in ASTM E 1461-01. Thermal diffusivity measurements were performed using LFA 467 equipment, supplied by NETZSCH, and conducted using the light flash technique. Temperature-modulated differential scanning calorimeter (TMDSC) technique was performed in a DSC 214 Polyma, supplied by NETZSCH, to determine the isobaric specific heat. The program established to perform isobaric specific heat measurements has been described previously [17]. Density values were determined by pycnometry. In addition, dynamic viscosity was studied due to the fact of its influence in the heat transfer process [20–22]. Dynamic viscosity measurements were obtained using a vibrational viscometer, model SV-10, supplied by A&D Company Ltd. The aforementioned thermal and rheological properties of nanofluids and HTF were analyzed in a temperature range from 290 K to 370 K.

3. Results

3.1. Morphological and Size Characterization of WS_2 and MoS_2 Nanostructures

Although MoS_2 and WS_2 are metal sulfides with similar properties, and the nanofluids were prepared with the same methodology, the TEM images revealed unexpected morphological results. Figure 1a shows the nanostructures found in the WS_2 nanofluid after LPE. We can observe nanosheets with lateral dimensions between 45 and 70 nm. However, in the MoS_2 nanofluid, the nanostructures obtained were nanowires as observed in Figure 1b. This is a surprising result, since the LPE process is typically used to obtain nanosheets from bulk material [23–26]. The nanowires obtained presented a diameter of approximately 27 nm and a 700 nm length (see Figure 1b). Inevitably, this morphological difference influenced the rest of the studied properties of nanofluids. On the other hand, the XRD patterns were registered for the MoS_2 and WS_2 solids extracted from the nanofluid after the LPE process. Figure 2 shows the patterns obtained. In both cases, we can observe that the pattern was typical for the hexagonal space group P63/mmc. The assignation of the planes according to the references (JCPDS cards No. 37-1492 and No. 08-0237 for MoS_2 and WS_2, respectively) is observed in the patterns shown in Figure 2.

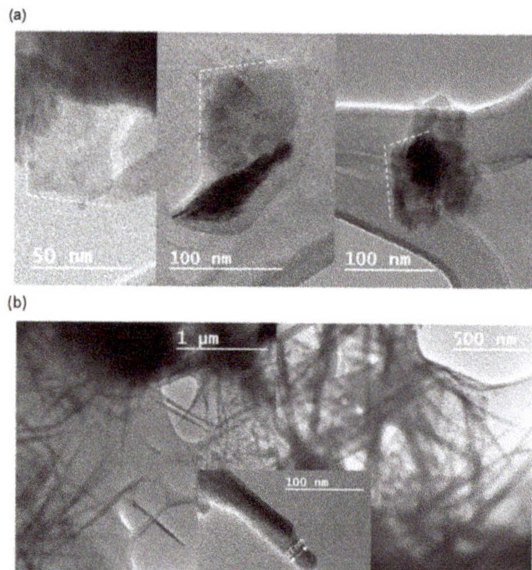

Figure 1. TEM images of WS_2 nanosheets (**a**) and MoS_2 nanowires (**b**) obtained after the liquid phase exfoliation (LPE) process.

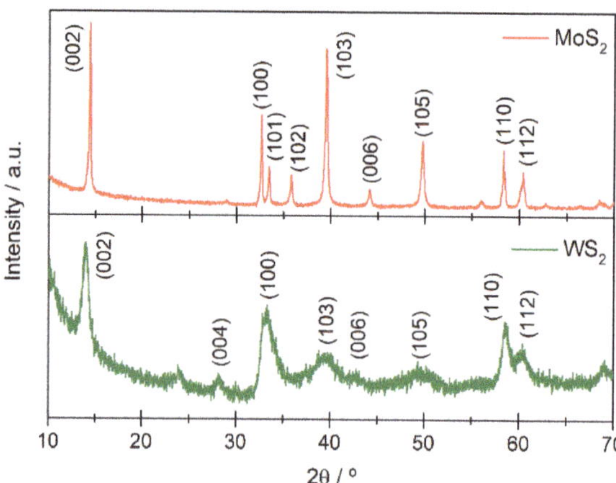

Figure 2. XRD patterns of MoS_2 nanowires and WS_2 nanosheets obtained after the LPE process.

3.2. Stability of Nanofluids

The stability study showed low levels of sedimentation over time for both nanofluids according to DLS and UV-Vis measurements. Figure 3a shows that, after three days of nanofluids preparation, the extinction coefficient was constant and, therefore, nanofluids began to stabilize. At the end of the characterization, the extinction coefficient at 678 nm in MoS_2 nanofluid decreased by 13%. For the WS_2 nanofluid, the decrease in extinction coefficient at 629 nm was 22%. As can be seen, the WS_2 nanofluid exhibited a higher extinction coefficient which suggests that a higher amount of nanomaterial was exfoliated during the LPE process. These results are concordant with those obtained by DLS technique (Figure 3b). The average hydrodynamic particle size found in WS_2 nanofluids was 200 nm, and it remained stable over the characterization time. Otherwise, in MoS_2 nanofluid, the average hydrodynamic particle size increased from 350 nm to 530 nm. Thus, in WS_2 nanofluid, the smaller particle size over time revealed that the agglomeration phenomenon and sedimentation process were not as prominent as in the MoS_2 nanofluid which led to a greater concentration of WS_2 nanostructures dispersed in HTF. In the case of MoS_2 nanofluids, the larger particle size would explain the lower extinction coefficient values of nanostructures present in the nanofluid compared to the WS_2 nanofluid, notwithstanding that at the end of the characterization, the agglomerates were not large enough for sedimentation to occur, since the extinction coefficient remained constant. Also, Figure 4 shows the UV-Vis spectra registered for both nanofluids right after preparation and seven days later in order to analyze shifts in the typical peaks observed for MoS_2 and WS_2. We can observe a decrease in the intensity of the spectra, but no significant shift of the absorption bands can be observed.

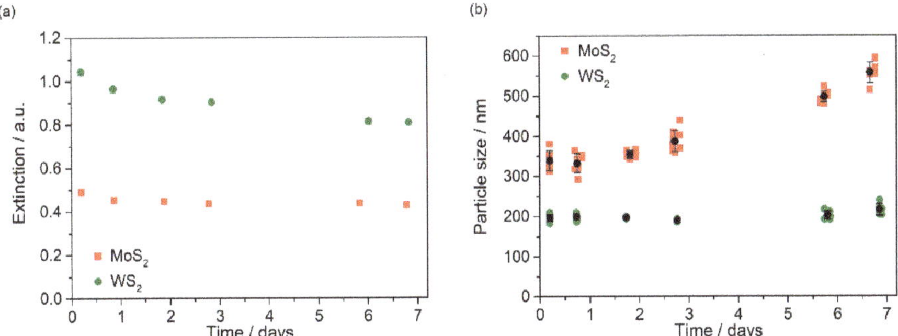

Figure 3. Extinction coefficient evolution obtained from UV-Vis spectra (**a**) and particle size results obtained from the DLS technique (**b**) for seven days.

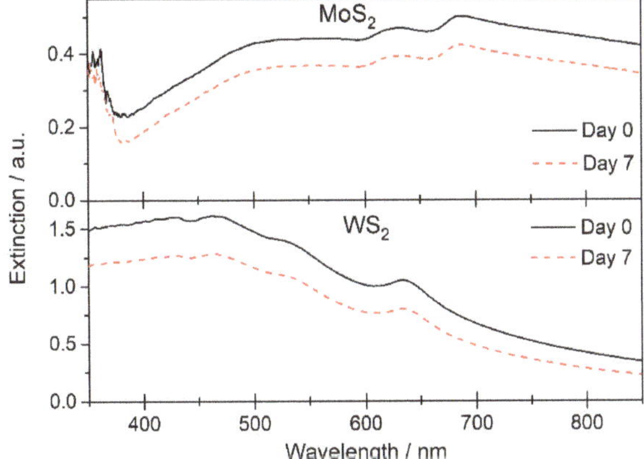

Figure 4. UV-Vis spectra for MoS_2- and WS_2-based nanofluids right after preparation and seven days later.

3.3. Nanofluid Performance

Some properties of nanofluids show a clear influence on heat transfer processes such as density, viscosity, isobaric specific heat, and thermal conductivity. In the case of density, an increase in density led to an enhancement in the heat transfer process. Also, it was expected that the nanofluid density was higher than the density of the base fluid, because density of liquids is lower than that of solids. As is observed in Table 1, WS_2 and MoS_2 nanofluids exhibited a slight increment in density values with respect to the pure HTF. This is a positive feature due to the high density values improving thermal conductivity [27–29]. In addition, the higher increase of density in WS_2 nanofluid was related to the higher nanostructure concentration. According to dynamic viscosity measurements, the values were close to that of HTF. The maximum increase in viscosity was 3.2% for MoS_2 nanofluid. However, these slight increases in viscosity were not significant enough to cause changes in pumping power or heat transfer.

Table 1. Density and dynamic viscosity results obtained for the base fluid and for the metal sulphide-based nanofluids at 298 K.

Sample	Density/kg m^{-3}	Dynamic viscosity /mPa s
HTF	1056.6 ± 0.5	3.70 ± 0.02
Nanofluid of MoS_2	1061.6 ± 0.2	3.82 ± 0.02
Nanofluid of WS_2	1069.3 ± 0.7	3.77 ± 0.01

Figure 5 shows the isobaric specific heat values obtained for nanofluids and base fluid. The highest increase of isobaric specific heat was up to 4.7% for MoS_2 nanofluid and up to 1.2% for WS_2 nanofluid with respect to HTF at 363 K. These results suggest that there was not a noticeable increase of this property which is understandable taking into account that isobaric specific heat of solids is lower than that of liquids. Furthermore, classic models based on mixture theory predict a decrease of isobaric-specific heat in respect to base fluid [30–32]. Notwithstanding, some recent theories do not consider nanofluids as a mixture of solid and liquid to determine the isobaric specific heat of nanofluids. These theories also include the study of the interaction between the molecules of nanostructure, surfactant, and liquid as a key factor in the enhancement of thermal resistance in the nanostructure/liquid interface [33–36]. Thus, these recent studies would explain the increase of isobaric-specific heat of nanofluids in respect to base fluid as it is reported in this paper and in previous works [37,38].

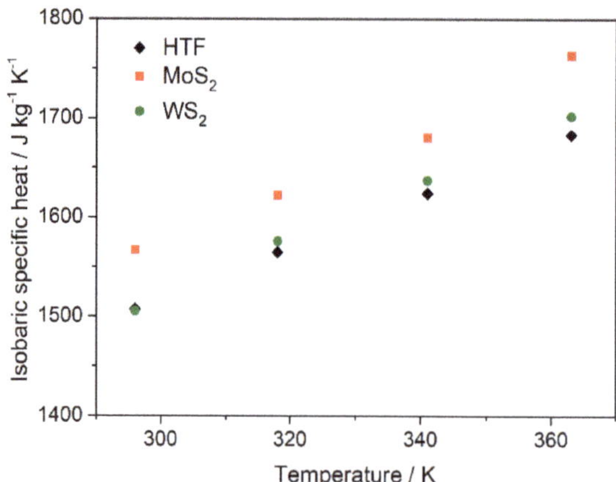

Figure 5. Isobaric-specific heat values of nanofluids and HTF.

Finally, the thermal conductivity of nanofluids and HTF was determined indirectly from the formula $k(T) = D(T)\cdot C_P(T)\cdot \rho(T)$ which considers thermal diffusivity (D), isobaric specific heat (C_P), and density (ρ). Figure 6a shows the thermal conductivity values of the nanofluids and the pure HTF for comparison. We can observe a different trend with temperature for nanofluids; that is, the thermal conductivity increases with temperature for nanofluids, while it decreases for the pure HTF. This opposite trend is favorable for the use of nanofluids in CSP plants, where high temperatures are used in operating conditions. This can be explained because, at high temperature, the collisions among nanostructures and between nanostructures and HTF molecules are intensified which leads to an increase in the Brownian motion and contributes to thermal conductivity enhancement. Figure 6b shows the thermal conductivity enhancement (TCE) calculated according to $TCE(\%) = \left(\frac{(k_{nf}-k_{bf})}{k_{bf}}\right) \times 100$.

A maximum increase of 45.6% for the nanofluid based on MoS_2 nanowires and 34.5% for the nanofluid based on WS_2 at 363 K were observed. Here, although bulk solids present similar thermal conductivities, as are shown above, the increase in the thermal conductivity values for nanofluids were different. Particularly interesting is that nanofluids based on MoS_2 with less concentration of nanostructure dispersed on HTF presented the largest thermal conductivity enhancement. A plausible reason for this phenomenon lies in the size and morphology of the nanostructures present in nanofluids. As observed during the DLS study, the stable agglomerates of MoS_2 nanowires were larger than the WS_2 nanosheet agglomerates. This greater agglomeration of nanowires in MoS_2 nanofluid can lead to percolation pathways that improve thermal conductivity more significantly than the well-distributed WS_2 nanosheets.

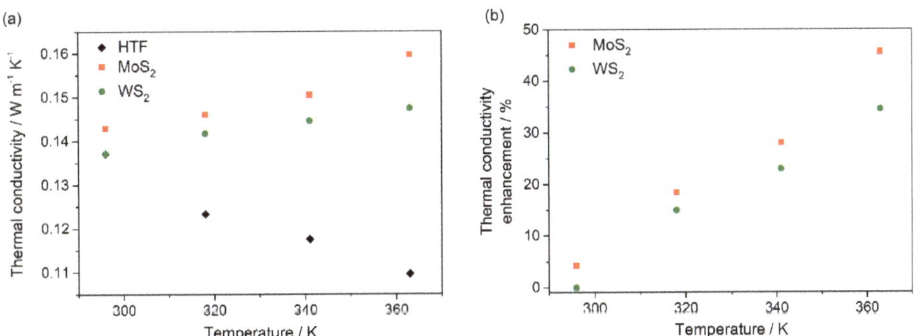

Figure 6. Thermal conductivity of base fluid and nanofluids (**a**) and the enhancement of the thermal conductivity for nanofluids (**b**).

In order to assess the efficiency of nanofluids in a CSP plant, the temperature of nanofluids and HTF at the heat pipe outlet was estimated. This temperature should be higher in nanofluids than in HTF if the heat transfer is improved. Mathematically, the heat flux between the surface of the pipe and the fluid, q_s'', is expressed as $q_s'' = h\Delta T = h(T_s - T_{m,0})$, where h is the heat transfer coefficient, T_s is the pipe surface temperature, and $T_{m,0}$ is the temperature of the fluid at the pipe outlet [39]. In this equation, for a constant solar irradiance of 1000 W m^{-2}, q_s'' and T_s are considered, and the ΔT for both the base fluid and for the nanofluids can be compared ($\Delta T_{nf}/\Delta T_{bf} = h_{bf}/h_{nf}$). In addition, the heat transfer coefficient can be estimated following the Mouromtseff number: $Mo = h = \left(\rho^{0.8}k^{0.67}C_P^{0.33}\right)/\mu^{0.47}$. Thus, Figure 7 shows the ratio of the Mouromtseff numbers of the nanofluids and the base fluid, which gives an idea of the enhancement of the heat transfer coefficient of the nanofluids analyzed. An increase up to 29% was observed. Finally, if $\Delta T_{nf}/\Delta T_{bf}1$, the outlet temperature $T_{m,0}$ was higher when a nanofluid was used with respect to HTF, and the efficiency of the collectors was improved. Figure 8 shows that MoS_2 and WS_2 nanofluids present higher outlet temperatures than HTF, since $\Delta T_{nf}/\Delta T_{bf}$ was lower than 1 in both cases. By increasing the temperature in the pipe, the difference in the outlet temperature of the nanofluid was accentuated with respect to HTF, improving the efficiency of the collector up to 21.3% at 363 K. The greater thermal conductivity and isobaric-specific heat of the nanofluid based on MoS_2 nanowires contributed to the outlet temperature rising up to 5% more than in the nanofluid based on WS_2 nanosheets.

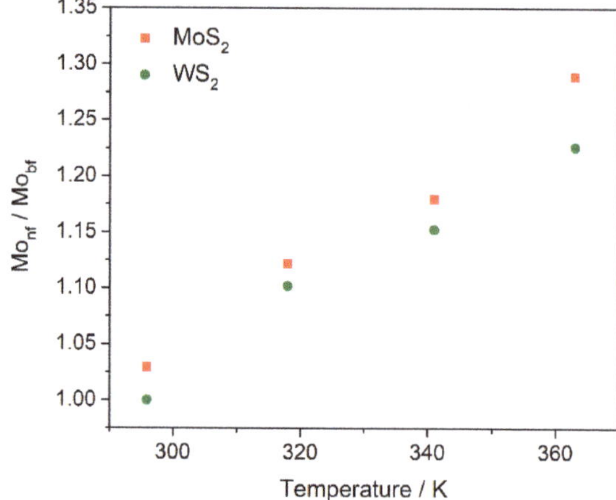

Figure 7. Ratio of the Mouromtseff numbers of the nanofluids prepared and the base fluid.

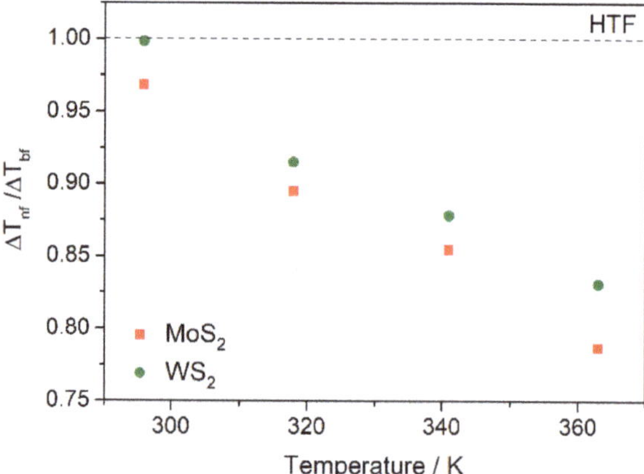

Figure 8. Comparison of the outlet temperature in solar collectors when metal sulfide nanofluids are used.

4. Conclusions

The LPE process was used in this work to prepare nanofluids based on WS_2 and MoS_2 nanostructures for their application as heat transfer fluids in CSP plants based on parabolic trough collectors. Thus, the base fluid used for nanofluids preparation was the eutectic mixture of diphenyl oxide and biphenyl commonly found in CSP plants.

The most striking result was that WS_2 and MoS_2 exhibited different behaviors during the same exfoliation process, and two different types of nanostructures were obtained: MoS_2 nanowires and WS_2 nanosheets. According to DLS and UV-Vis spectroscopy, nanofluids based on MoS_2 nanowires and WS_2 nanosheets presented a high stability over time which suggests that this kind of advanced nanomaterials can be an alternative to nanofluids based on spherical nanoparticles. In the nanofluid based on WS_2 nanosheets, the concentration of nanomaterial was higher, and agglomerates were smaller than in the nanofluid based on MoS_2 nanowires. Nevertheless, nanofluid based on MoS_2 nanowires exhibited a greater improvement of thermal properties despite the lower concentration of nanostructures dispersed in HTF compared to the WS_2 nanofluid and the similar thermal conductivity of the metal chalcogenides. The larger agglomerates of MoS_2 nanowires led to percolation paths through which heat was transferred faster and more efficiently than through the well-distributed WS_2 nanosheets.

Notwithstanding, both nanofluids presented significant improvements in the heat transfer process with respect to HTF. Thus, the thermal conductivity increased up to 45.6% in the nanofluid based on MoS_2 nanowires and 34.5% in the nanofluid based on WS_2. In addition, isobaric-specific heat was measured and a slight increase of up to 4.7% for MoS_2 nanofluid and 1.2% for WS_2 nanofluid were obtained which is consistent with the current isobaric-specific models. Finally, the nanofluid temperatures at the outlet of the absorber tube used in CSP plants were estimated and compared with that obtained for the pure HTF. The results revealed that the efficiency of the collector was increased up to 21.3% and 16.8% when the MoS_2 or WS_2 nanofluids were used, respectively. Therefore, nanofluids based on metal chalcogenides show properties of great interest to be used in CSP plants based on parabolic trough collectors, such as their high stability and their thermophysical properties improvements, compared to those of the heat transfer fluid currently used.

Author Contributions: Methodology, P.M.-M., T.A., J.J.G., I.C.-B., R.G.-V.; investigation, P.M.-M., R.A., M.R.-F.; writing—original draft preparation, P.M.-M.; writing—review and editing, J.N.; supervision, J.N.; project administration, J.N.; funding acquisition, J.N.

Funding: This research was funded by Ministerio de Ciencias, Innovación y Universidades (Spanish Government), grant numbers RTI2018-096393-B-I00 and UNCA15-CE-2945. The APC was funded by Nanouptake COST Action CA15119.

Acknowledgments: This investigation was a contribution to the COST (European Cooperation in Science and Technology) Action CA15119: Overcoming Barriers to Nanofluids Market Uptake (NanoUptake). P.M.-M. acknowledges the EU COST Action CA15119: Overcoming Barriers to Nanofluids Market Uptake for financial support in the participation of the 1st International Conference on Nanofluids (ICNf) and the 2nd European Symposium on Nanofluids (ESNf) participation.

Conflicts of Interest: The authors declare no conflict of interest.

References

1. Desideri, U.; Campana, P.E. Analysis and comparison between a concentrating solar and a photovoltaic power plant. *Appl. Energy* **2014**, *113*, 422–433. [CrossRef]
2. Xu, B.; Li, P.W.; Chan, C. Application of phase change materials for thermal Energyy storage in concentrated solar thermal power plants: A review to recent developments. *Appl. Energy* **2015**, *160*, 286–307. [CrossRef]
3. Rovense, F.; Reyes-Belmonte, M.A.; Gonzalez-Aguilar, J.; Amelio, M.; Bova, S.; Romero, M. Flexible electricity dispatch for CSP plant using un-fired closed air Brayton cycle with particles based thermal Energyy storage system. *Energy* **2019**, *173*, 971–984. [CrossRef]
4. Zhang, H.L.; Baeyens, J.; Degreve, J.; Caceres, G. Concentrated solar power plants: Review and design methodology. *Renew. Sust. Energy Rev.* **2013**, *22*, 466–481. [CrossRef]
5. Islam, M.T.; Huda, N.; Abdullah, A.B.; Saidur, R. A comprehensive review of state-of-the-art concentrating solar power (CSP) technologies: Current status and research trends. *Renew. Sust. Energy Rev.* **2018**, *91*, 987–1018. [CrossRef]
6. Liu, M.; Tay, N.H.S.; Bell, S.; Belusko, M.; Jacob, R.; Will, G.; Saman, W.; Bruno, F. Review on concentrating solar power plants and new developments in high temperature thermal Energyy storage technologies. *Renew. Sust. Energy Rev.* **2016**, *53*, 1411–1432. [CrossRef]
7. Wang, F.Q.; Cheng, Z.M.; Tan, J.Y.; Yuan, Y.; Shuai, Y.; Liu, L.H. Progress in concentrated solar power technology with parabolic trough collector system: A comprehensive review. *Renew. Sust. Energy Rev.* **2017**, *79*, 1314–1328.
8. Fernandez-Garcia, A.; Zarza, E.; Valenzuela, L.; Perez, M. Parabolic-trough solar collectors and their applications. *Renew. Sust. Energy Rev.* **2010**, *14*, 1695–1721. [CrossRef]
9. Price, H.; Lupfert, E.; Kearney, D.; Zarza, E.; Cohen, G.; Gee, R.; Mahoney, R. Advances in parabolic trough solar power technology. *J. Sol. Energy T Asme* **2002**, *124*, 109–125. [CrossRef]
10. Morin, G.; Dersch, J.; Platzer, W.; Eck, M.; Haberle, A. Comparison of Linear Fresnel and Parabolic Trough Collector power plants. *Sol. Energy* **2012**, *86*, 1–12. [CrossRef]

11. Tyagi, H.; Phelan, P.; Prasher, R. Predicted Efficiency of a Low-Temperature Nanofluid-Based Direct Absorption Solar Collector. *J. Sol. Energy Eng.* **2009**, *131*, 041004. [CrossRef]
12. Taylor, R.A.; Phelan, P.E.; Otanicar, T.P.; Walker, C.A.; Nguyen, M.; Trimble, S.; Prasher, R. Applicability of nanofluids in high flux solar collectors. *J. Renew. Sustain. Energy* **2011**, *3*, 023104. [CrossRef]
13. Mwesigye, A.; Huan, Z.J.; Meyer, J.P. Thermodynamic optimisation of the performance of a parabolic trough receiver using synthetic oil-Al2O3 nanofluid. *Appl. Energy* **2015**, *156*, 398–412. [CrossRef]
14. Peng, B.; Zhang, H.; Shao, H.Z.; Xu, Y.C.; Zhang, X.C.; Zhu, H.Y. Thermal conductivity of monolayer MoS2, MoSe2, and WS2: Interplay of mass effect, interatomic bonding and anharmonicity. *RSC Adv.* **2016**, *6*, 5767–5773. [CrossRef]
15. Peimyoo, N.; Shang, J.Z.; Yang, W.H.; Wang, Y.L.; Cong, C.X.; Yu, T. Thermal conductivity determination of suspended mono- and bilayer WS2 by Raman spectroscopy. *Nano Res.* **2015**, *8*, 1210–1221. [CrossRef]
16. Yan, R.S.; Simpson, J.R.; Bertolazzi, S.; Brivio, J.; Watson, M.; Wu, X.F.; Kis, A.; Luo, T.F.; Walker, A.R.H.; Xing, H.G. Thermal Conductivity of Monolayer Molybdenum Disulfide Obtained from Temperature-Dependent Raman Spectroscopy. *ACS Nano* **2014**, *8*, 986–993. [CrossRef]
17. Navas, J.; Sanchez-Coronilla, A.; Martin, E.I.; Teruel, M.; Gallardo, J.J.; Aguilar, T.; Gomez-Villarejo, R.; Alcantara, R.; Fernandez-Lorenzo, C.; Pinero, J.C.; et al. On the enhancement of heat transfer fluid for concentrating solar power using Cu and Ni nanofluids: An experimental and molecular dynamics study. *Nano Energy* **2016**, *27*, 213–224. [CrossRef]
18. Mishra, A.K.; Lakshmi, K.V.; Huang, L.P. Eco-friendly synthesis of metal dichalcogenides nanosheets and their environmental remediation potential driven by visible light. *Sci. Rep. UK* **2015**, *5*, 15718. [CrossRef]
19. Song, H.L.; Yu, X.F.; Chen, M.; Qiao, M.; Wang, T.J.; Zhang, J.; Liu, Y.; Liu, P.; Wang, X.L. Modification of WS2 nanosheets with controllable layers via oxygen ion irradiation. *Appl. Surf. Sci.* **2018**, *439*, 240–245. [CrossRef]
20. Murshed, S.M.S.; Leong, K.C.; Yang, C. Investigations of thermal conductivity and viscosity of nanofluids. *Int. J. Therm. Sci.* **2008**, *47*, 560–568. [CrossRef]
21. Prasher, R.; Song, D.; Wang, J.L.; Phelan, P. Measurements of nanofluid viscosity and its implications for thermal applications. *Appl. Phys. Lett.* **2006**, *89*, 133108. [CrossRef]
22. Khanafer, K.; Vafai, K. A critical synthesis of thermophysical characteristics of nanofluids. *Int. J. Heat Mass Trans.* **2011**, *54*, 4410–4428. [CrossRef]
23. Shen, J.F.; He, Y.M.; Wu, J.J.; Gao, C.T.; Keyshar, K.; Zhang, X.; Yang, Y.C.; Ye, M.X.; Vajtai, R.; Lou, J.; et al. Liquid Phase Exfoliation of Two-Dimensional Materials by Directly Probing and Matching Surface Tension Components. *Nano Lett.* **2015**, *15*, 5449–5454. [CrossRef] [PubMed]
24. Teruel, M.; Aguilar, T.; Martinez-Merino, P.; Carrillo-Berdugo, I.; Gallardo-Bernal, J.J.; Gomez-Villarejo, R.; Alcantara, R.; Fernandez-Lorenzo, C.; Navas, J. 2D MoSe2-based nanofluids prepared by liquid phase exfoliation for heat transfer applications in concentrating solar power. *Sol. Energy Mat. Sol. Cells* **2019**, *200*, 109972. [CrossRef]
25. Lotya, M.; Hernandez, Y.; King, P.J.; Smith, R.J.; Nicolosi, V.; Karlsson, L.S.; Blighe, F.M.; De, S.; Wang, Z.M.; McGovern, I.T.; et al. Liquid Phase Production of Graphene by Exfoliation of Graphite in Surfactant/Water Solutions. *J. Am. Chem. Soc.* **2009**, *131*, 3611–3620. [CrossRef]
26. Nicolosi, V.; Chhowalla, M.; Kanatzidis, M.G.; Strano, M.S.; Coleman, J.N. Liquid Exfoliation of Layered Materials. *Science* **2013**, *340*, 1226419. [CrossRef]
27. Keblinski, P.; Eastman, J.A.; Cahill, D.G. Nanofluids for thermal transport. *Mater. Today* **2005**, *8*, 36–44. [CrossRef]
28. Timofeeva, E.V.; Yu, W.H.; France, D.M.; Singh, D.; Routbort, J.L. Nanofluids for heat transfer: An engineering approach. *Nanoscale Res. Lett.* **2011**, *6*, 182. [CrossRef]
29. Mohammed, H.A.; Hasan, H.A.; Wahid, M.A. Heat transfer enhancement of nanofluids in a double pipe heat exchanger with louvered strip inserts. *Int. Commun. Heat Mass* **2013**, *40*, 36–46. [CrossRef]
30. Taylor, R.; Coulombe, S.; Otanicar, T.; Phelan, P.; Gunawan, A.; Lv, W.; Rosengarten, G.; Prasher, R.; Tyagi, H. Critical Review of the Novel Applications and Uses of Nanofluids. In Proceedings of the Asme Micro/Nanoscale Heat and Mass Transfer International Conference, Atlanta, GA, USA, 3–6 March 2012; pp. 219–234.
31. Zhou, S.Q.; Ni, R. Measurement of the specific heat capacity of water-based Al_2O_3 nanofluid. *Appl. Phys. Lett.* **2008**, *92*, 093123. [CrossRef]

32. Vajjha, R.S.; Das, D.K. Specific Heat Measurement of Three Nanofluids and Development of New Correlations. *J. Heat Transf.* **2009**, *131*, 071601. [CrossRef]
33. Li, L.; Zhang, Y.W.; Ma, H.B.; Yang, M. Molecular dynamics simulation of effect of liquid layering around the nanoparticle on the enhanced thermal conductivity of nanofluids. *J. Nanopart. Res.* **2010**, *12*, 811–821. [CrossRef]
34. Oh, S.H.; Kauffmann, Y.; Scheu, C.; Kaplan, W.D.; Ruhle, M. Ordered liquid aluminum at the interface with sapphire. *Science* **2005**, *310*, 661–663. [CrossRef]
35. Prasher, R.; Bhattacharya, P.; Phelan, P.E. Brownian-motion-based convective-conductive model for the effective thermal conductivity of nanofluids. *J. Heat Transf.* **2006**, *128*, 588–595. [CrossRef]
36. Xue, L.; Keblinski, P.; Phillpot, S.R.; Choi, S.U.S.; Eastman, J.A. Effect of liquid layering at the liquid-solid interface on thermal transport. *Int. J. Heat Mass Trans.* **2004**, *47*, 4277–4284. [CrossRef]
37. Starace, A.K.; Gomez, J.C.; Wang, J.; Pradhan, S.; Glatzmaier, G.C. Nanofluid heat capacities. *J. Appl. Phys.* **2011**, *110*, 124323. [CrossRef]
38. De Castro, C.A.N.; Murshed, S.M.S.; Lourenco, M.J.V.; Santos, F.J.V.; Lopes, M.L.M.; Franca, J.M.P. Enhanced thermal conductivity and specific heat capacity of carbon nanotubes ionanofluids. *Int. J. Therm. Sci.* **2012**, *62*, 34–39. [CrossRef]
39. Zare, V.; Moalemian, A. Parabolic trough solar collectors integrated with a Kalina cycle for high temperature applications: Energyy, exergy and economic analyses. *Energy Convers. Manag.* **2017**, *151*, 681–692. [CrossRef]

Article

Computational Study of Flow and Heat Transfer Characteristics of EG-Si_3N_4 Nanofluid in Laminar Flow in a Pipe in Forced Convection Regime †

Edin Berberović [1,*] and Siniša Bikić [2]

1 Polytechnic Faculty, University of Zenica, Fakultetska 1, 72000 Zenica, Bosnia and Herzegovina
2 Faculty of Technical Sciences, University of Novi Sad, Trg Dositeja Obradovića 6, 21000 Novi Sad, Serbia; bika@uns.ac.rs
* Correspondence: eberberovic@ptf.unze.ba; Tel.: +387-32-449-120

† This paper is an extended version of our paper published in the 1st International Conference on Nanofluids (ICNf) and 2nd European Symposium on Nanofluids (ESNf), 26–28 June 2019, Castelló, Spain. Conference Proceedings. pp. 226–230.

Received: 12 November 2019; Accepted: 9 December 2019; Published: 22 December 2019

Abstract: Laminar flow of ethylene glycol-based silicon nitride (EG-Si_3N_4) nanofluid in a smooth horizontal pipe subjected to forced heat convection with constant wall heat flux is computationally modeled and analyzed. Heat transfer is evaluated in terms of Nusselt number (Nu) and heat transfer coefficient for various volume fractions of Si_3N_4 nanoparticles in the base fluid and different laminar flow rates. The thermophysical properties of the EG-Si_3N_4 nanofluid are taken from a recently published experimental study. Computational modelling and simulation are performed using open-source software utilizing finite volume numerical methodology. The nanofluid exhibits non-Newtonian rheology and it is modelled as a homogeneous single-phase mixture, the properties of which are determined by the nanoparticle volume fraction. The existing features of the software to simulate single-phase flow are extended by implementing the energy transport coupled to the fluid flow and the interaction of the fluid flow with the surrounding pipe wall via the applied wall heat flux. In addition, the functional dependencies of the thermophysical properties of the nanofluid on the volume fraction of nanoparticles are implemented in the software, while the non-Newtonian rheological behavior of the nanofluid under consideration is also taken into account. The obtained results from the numerical simulations show very good predicting capabilities of the implemented computational model for the laminar flow coupled to the forced convection heat transfer. Moreover, the analysis of the computational results for the nanofluid reflects the increase of heat transfer of the EG-Si_3N_4 nanofluid in comparison to the EG for all the considered nanoparticle volume fractions and flow rates, indicating promising features of this nanofluid in heat transfer applications.

Keywords: EG-Si_3N_4 nanofluid; numerical simulation; laminar flow; heat transfer coefficient

1. Introduction

Nanofluids are mixtures or suspensions of small particles of order of tens of nanometer in pure base fluids, which after mixing and dispersing the nanoparticles have shown the potential to increase heat transfer in engineering applications. The increase in thermal conductivity of nanofluids relative to the base fluid was first time indicated by Choi and Eastman [1]. They conducted a theoretical study of thermal conductivity of nanofluids with copper nanophase materials. Consequently, due to their discovered potential in a variety of applications incorporating heat transfer, such as heat exchangers, solar collectors, or thermal energy storage systems, the research in nanofluids has been intensified over the past years [2–6]. For the base fluids water and ethylene glycol are commonly used in both, the experimental [5,7–14] and the numerical studies [4,15–17]. The numerical investigation has been directed towards two main different approaches by modeling the nanofluid flow either as an effective single-phase mixture with unique properties depending on the nanoparticle volume fraction, or a two-phase flow with explicitly accounting for the interaction of the nanoparticles with the base fluid [18]. In the last several years ionic liquids (ILs) have been used more and more as base fluids for producing the nanofluids [19–25]. At ambient conditions ionic liquids have the property of being non-volatile and non-flammable and they are recyclable liquids. These are the reasons why ionic liquids are very often considered as green fluids. By suspending nanoparticles in an ionic liquid, the so called ionanofluid (INF) is formed. Ionanofluids is a new type of nanofluids which can further increase the thermophysical properties of ionic liquids. In a recent study [26] the available research and results are summarized in order to assess the potential of ionanofluids. In spite of scattering of the literature results, it was concluded that ionanofluids have good potential in advanced thermal and energy applications. The recently conducted researches especially point at ionanofluids as potential media for advanced solar collectors [19–25].

In addition to the convincing results for the enhancement of the heat transfer a broader approach to research in nanofluids has to identify the drawback of their usage, such as the obvious problem of sedimentation of the nanoparticles after a long-time storage [27–29], or even the costs that have to be paid for the increased efficiency in the energy performance [30], since the enhancement in heat transfer characteristics might not be enough to balance the increase the pressure drop in terms of costs.

The materials of nanoparticles used for forming the nanofluids are commonly oxides, metals, carbides, and carbon nanotubes. In the last several years suspensions of nitride nanoparticles in base fluids have been started to be used for research [9–13,31–35]. Promising results have been obtained so far which point out that thermal conductivity enhances with the volume fraction of nanoparticles in suspension. The increase of the thermal conductivity for hexagonal boron nitride nanofluids (average particle size 70 nm) goes up to 26% at volume fractions of 3% for water and up to and 16% for the same volume fraction in EG-based nanofluids [10]. The research of thermophysical properties of EG-based nanofluids with TiN nanoparticles has been conducted for average nanoparticle diameters of 30 nm and 50 nm and nanoparticle volume fraction from 0.0022 to 0.0111. It was concluded that thermal conductivity of considered nanofluids could be improved by decreasing the size of nanoparticles [13]. Contrary to that, the thermal conductivity of nanofluids with boron nitride (BN) nanoparticles in ethylene glycol was found to be possible to be enhanced with increasing the nanoparticle size [9]. The main reason for this was concluded to be the surface area and the aspect ratio of BN nanoparticles. In the same study one more abnormality related to thermal conductivity enhancements of BN/EG nanofluids is noticed. The higher increase of the thermal conductivity was achieved at a low nanoparticle volume fraction. The authors suspect the chain-like loose aggregation of nanoparticles to be the reason for this uncommon increase of thermal conductivity at very low volume fraction of nanoparticles. Nanofluids with suspended aluminum nitride nanoparticles (AlN) in ethylene glycol (average particle sizes 165 nm) and propylene glycol (average particle sizes 169 nm) as base fluids with the nanoparticle volume fraction of 0.1 increase the thermal conductivity by 38.71% and 40.2%, respectively [31]. Computational fluid dynamics using the finite-volume method and adopting the SIMPLE algorithm was performed for flowing of suspension of aluminum nitride nanoparticles (particle size 30 nm) in ethylene glycol as the base fluid through a horizontal heated tube [32]. Overall efficiency of heat transfer enhancement was above one for ranges of nanoparticle concentrations from 1% to 3% and Reynolds numbers from zero to 20,000. The heat capacity of nanofluids with ethylene glycol and three different types of nitride nanoparticles (namely, aluminum nitride AlN, silicon nitride Si_3N_4, and titanium nitride (TiN) was experimentally determined in [33]. It was concluded that the volume fraction of nanoparticles has a strong effect on the heat capacity, whereas the nanoparticle size does not show a significant impact on this thermophysical property.

The findings presented above nominate nanofluids with suspended nitrides for in-depth research in the future. In numerical simulations nanofluid is often modelled as a single-phase Newtonian fluid, which is considered to be a good balance between the achieved numerical accuracy and the acceptable computational costs, in particular for use in practical engineering applications, which may be accompanied by flows in complex geometries or involve some additional complex physics. However, in the available literature there is a lack in studies of nanofluids which exhibit some more complex rheology. The present study focuses on computational modelling and analysis of a laminar flow of a non-Newtonian nanofluid (ethylene glycol-based silicon nitride (EG-Si_3N_4)) in a horizontal pipe with respect to the forced convection heat transfer.

The nanofluid considered in the present study is ethylene glycol-based silicon nitride, which was experimentally examined in [35] by measuring its rheological behavior and thermal properties in addition to the electrical conductivity and optical properties, all as functions of the concentration of nanoparticles. In the above-mentioned experimental study, the average size of the nanoparticles is 20 nm and the suspension was produced in a production method using vortex shaker, ultrasonic bath, and high-energy ultrasound generator. The properties of the nanofluid are determined for volume fractions of nanoparticles up to 0.035. The obtained experimental results reveal the non-Newtonian shear thinning behavior of the nanofluid, a simple linear increase of thermal conductivity with volume fraction of nanoparticles, as well as a high enhancement of electrical conductivity and an increase of optical properties (refractive index, absorption) with volumetric concentration of nanoparticles. Such behavior of the properties renders the EG-Si_3N_4 nanofluid as being convenient not only in applications for heat transfer, such as heat exchangers and thermal energy storage, but also for applications where enhanced electrical and optical properties are of importance, such as electromagnetic flow meters or solar collectors.

The present computational study is devoted to determining the heat transfer characteristics of the EG-Si_3N_4 nanofluid when it is subjected to laminar pipe flow and forced convection heat transfer with a constant wall heat flux. Heat transfer is evaluated by determining the Nusselt number and the heat transfer coefficient for various volume fractions of Si_3N_4 nanoparticles at different nanofluid flow rates. The computational model is implemented in the open-source software foam-extend ver. 4.0 [36], which represents an extension of the widely used open-source software OpenFOAM® [37] utilizing finite volume numerical methodology. The results show that the heat transfer of the EG-Si_3N_4 nanofluid increases in comparison with the ethylene glycol as the base fluid.

2. Computational Model

2.1. Governing Transport Equations and Constitutive Relations

The nanofluid considered here is modeled as a single-phase mixture with thermophysical properties depending on the nanoparticle volume fraction. The computational model for the steady-state flow of such a fluid is given by the governing equations for the mass, linear momentum, and energy conservation with neglected temperature induced density changes:

$$\nabla \cdot (\rho \boldsymbol{U}) = 0, \tag{1}$$

$$\nabla \cdot (\rho \boldsymbol{U}\boldsymbol{U}) = -\nabla p + \nabla \cdot \tau, \tag{2}$$

$$\nabla \cdot \left(\rho c_p \boldsymbol{U} T\right) = \nabla \cdot (k \nabla T), \tag{3}$$

where ρ is the density of the nanofluid, $\boldsymbol{U}$ is the velocity, p is the pressure, τ is the viscous stress tensor, c_p is the specific heat, k is the thermal conductivity, and T is the temperature. The changes in density due to temperature changes in the fluid are neglected since the effects of the natural heat convection are not considered. This is a consequence of the assumption that the effects of the forced heat convection are predominant in comparison to the natural heat convection. This assumption is justified for cases where the flow is constrained to small geometries, such as the one used here (tube with 4 mm inner diameter), whereas the natural heat convection effects become important in larger geometries [38].

To evaluate the density and the specific heat of the nanofluid, weighted averages based on the volume fraction of nanoparticles are use, according to the expressions:

$$\rho = \varphi_v \rho_n + (1 - \varphi_v)\rho_{bf}, \tag{4}$$

$$\rho c_p = \varphi_v \left(\rho c_p\right)_n + (1 - \varphi_v)\left(\rho c_p\right)_{bf}, \tag{5}$$

where φ_v is the volume fraction of nanoparticles in the suspension and the indices n and bf refer to nanoparticles and base fluid, respectively.

The thermal conductivity of the EG-Si_3N_4 nanofluid is modelled as a linear dependency on the nanoparticle volume fraction, with the expressions [35]:

$$k = (1 + 2.87\varphi_v)k_{bf}. \tag{6}$$

The rheology of the EG-Si_3N_4 nanofluid is modelled with the expressions [35]:

$$\tau = \tau_0 + K\gamma^n, \tag{7}$$

where γ represents the strain rate, while the parameters τ_0, K, and n depend on the Si_3N_4 nanoparticle volume fraction φ_v and their values are provided in [35]. The Expression (7) represents the Herschel-Bulkley rheology model, which is implemented in the software foam-extend in a manner that avoids the numerical stability problems for very low values of the strain rate and the singularity at the limiting case of $\gamma = 0$, which would mathematically correspond to infinite viscosity. The local viscous stress tensor can be represented as the product of the local nanofluid dynamic viscosity μ and the local strain rate γ, corresponding to the given local viscous stress. If the dynamic viscosity is expressed in terms of the kinematic viscosity, the local viscous stress is $\tau = \rho\nu\gamma$ and the Equation (2) can be rewritten in the following form:

$$\nabla \cdot (\boldsymbol{UU}) = -\frac{1}{\rho}\nabla p + \frac{1}{\rho}\nabla \cdot \tau = -\frac{1}{\rho}\nabla p + \frac{1}{\rho}\nabla \cdot (\mu\gamma) = -\frac{1}{\rho}\nabla p + \frac{1}{\rho}\nabla \cdot (\rho\nu\gamma). \tag{8}$$

The kinematic viscosity is determined from the expression:

$$\nu = \frac{1}{\rho}\frac{\tau}{\gamma} = \frac{1}{\rho}\frac{\tau_0 + K\left[\gamma^n - \left(\frac{\tau_0}{\nu_0}\right)^n\right]}{\dot{\gamma}} = \frac{1}{\rho}\frac{\tau_0 + K\left(\gamma^n - \gamma_0^n\right)}{\gamma}, \tag{9}$$

where the small value for the strain rate $\gamma_0 = \tau_0/\nu_0$ is introduced in order to be able to control the situation of very small local strain rates during the computation. Thus, the viscous stress is evaluated from the expression:

$$\tau = \tau_0 + K\left(\gamma^n - \gamma_0^n\right), \tag{10}$$

and the graphical representation is given in Figure 1. The value for γ_0 is determined by setting the viscosity ν_0 to some high value (say 0.1), as the input parameter to the model. The value for the viscosity, which is required in the momentum equation, is then determined as being equal to ν_0 for local values of $\gamma < \gamma_0$ and the viscous stress is then closely equal to τ_0, whereas for $\gamma > \gamma_0$ the viscous stress calculated from the expression (10) will nearly be equal to the theoretical value in the expression (7). In this manner, the singularity at $\gamma = \gamma_0$ is avoided and the numerically obtained viscous stress is negligibly different from the theoretical value in the expression (7), the difference being determined by the value for ν_0 and γ_0: the higher the value for ν_0 is set, the lower will be the value for γ_0 and the viscous stress will be closer to the theoretical value.

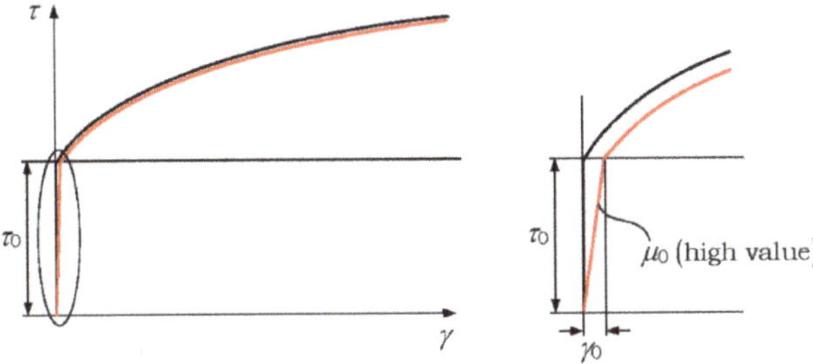

Figure 1. Herschel-Bulkley model for non-Newtonian fluid rheology.

2.2. Geomery Modeling and Numerical Mesh

The geometry consists of a smooth horizontal pipe in which the flow is axisymmetric due to the assumption of negligible effects of natural heat convection. Therefore, in order to significantly reduce the computational efforts, the pipe can be modelled as a 2-D axisymmetric slice and the numerical mesh is two-dimensional having only one cell in the azimuthal direction, as shown in Figure 2.

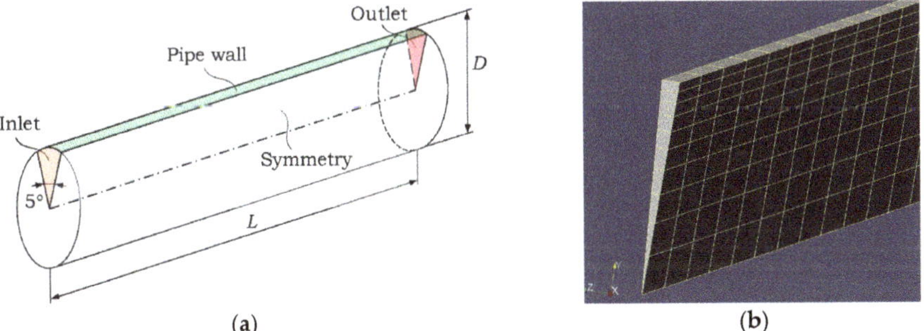

Figure 2. Geometry and mesh modeling: (**a**) 2-D axisymmetric domain flow; (**b**) Snapshot of the 2-D numerical mesh.

2.3. Discretization of Transport Equations and Boundary Conditions

For the discretization of various terms in the governing transport equations the finite-volume numerical methodology is used. The numerical mesh is made of a number of small volumes (cells), in which each two cells share one flat cell-face. All variables are stored at cell-centroids (cell-centers) denoted with P and all cells are bounded by a finite number of flat cell-faces denoted with f. The transport equations representing conservation laws in a steady state simulation may be written in a generic form for some variable ϕ:

$$\nabla \cdot (\rho \boldsymbol{U} \phi) = \nabla \cdot (\Gamma \nabla \phi) + S_{\phi}(\phi), \tag{11}$$

where ϕ stands for one in the continuity equation, for $\boldsymbol{U}$ in the momentum equation and for T in the energy equation, Γ is the diffusion coefficient, replaced by v in the momentum equation and by k/c_p in the energy equation, and S_{ϕ} represent any sources which may be present and depend on ϕ. There is no such source in the present model in the energy equation, so the only term treated as a source of momentum is the negative pressure gradient in the momentum equation. All terms in Equation (11) are discretized within the finite-volume approximation and the transport equations are integrated over all cell-volumes:

$$\int_{V_P} \nabla \cdot (\rho \boldsymbol{U} \phi) \mathrm{d}V = \int_{V_P} \nabla \cdot (\Gamma \nabla \phi) \mathrm{d}V + \int_{V_P} S_{\phi}(\phi) \mathrm{d}V \tag{12}$$

The discretization of terms containing spatial derivatives is performed by converting the volume integrals in surface integrals according to Gauss's theorem and then summing over all cell-faces f is performed. The gradient-containing terms are approximated according to the expression:

$$\int_{V_P} \nabla\phi \mathrm{d}V = \int_{S_P} \phi \mathrm{d}\mathbf{S} \approx \sum_f \mathbf{S}_f \phi_f. \tag{13}$$

where S_P is the surface area of all the cell-faces enclosing the control volume V_P, $\mathbf{S}_f$ is the surface of a cell-face, and $\mathrm{d}\mathbf{S}$ is the differential of the cell-face surface-normal vector. The summation is again done over all cell-faces f bounding a cell P. In accordance, divergence-containing terms are discretized as:

$$\int_{V_P} \nabla \cdot \phi \mathrm{d}V = \int_{S_P} \phi \cdot \mathrm{d}\mathbf{S} \approx \sum_f \mathbf{S}_f \cdot \phi_f \tag{14}$$

Similarly, the second-order spatial terms are approximated as:

$$\int_{V_P} \nabla \cdot (\Gamma \nabla \phi) \mathrm{d}V = \int_{S_P} \mathrm{d}\mathbf{S} \cdot (\Gamma \nabla \phi) \approx \sum_f \Gamma_f \mathbf{S}_f \cdot \left(\nabla \phi_f\right). \tag{15}$$

For the calculation of surface integrals, the unknown variables are interpolated to the centers of the cell-faces. The approximation of the terms with gradients does not impose severe restrictions, so that simple linear interpolation can be used, while on the other hand the discretization of the term involving the divergence, in particular the convective term in the momentum equation, is more problematic and approximation of the convective term represents one of the challenges in Computational Fluid Dynamics (CFD). Due to the simplicity of the problem studied a simple upwind convection scheme is considered to be sufficient here for the convective terms in both, the momentum and the energy equations. In the momentum equation the convective term is linearized by calculating the volume flux through the cell-faces from the previous iteration.

In approximations of source terms, the corresponding values at cell-centers are reconstructed from the values at the cell-faces. For example, the source containing the pressure gradient is calculated according to the expression:

$$(\nabla p)_P \approx \left(\sum_f \frac{\mathbf{S}_f \mathbf{S}_f}{|\mathbf{S}_f|}\right)^{-1} \cdot \left(\sum_f \frac{\mathbf{S}_f}{|\mathbf{S}_f|} \cdot \left(\nabla p_f^{\perp}\right) |\mathbf{S}_f|\right), \tag{16}$$

where the symbol $\perp$ denotes the surface-normal gradient, which is obtained by a simple linear scheme between two neighboring computational cells.

For solving the coupled flow and heat transfer problem described by transport Equations (1)–(3) and the constitutive relations (4)–(10), boundary conditions are prescribed at domain boundaries (pipe inlet, pipe outlet, and pipe wall, with reference to Figure 2). The boundary conditions are summarized in the following.

Momentum:

- inlet: $\boldsymbol{U} = (U_{in}, 0, 0)$, $\nabla p = 0$,
- pipe wall: $\boldsymbol{U} = (0, 0, 0)$, $\nabla p = 0$,
- outlet: $\nabla \boldsymbol{U} = 0$, $p = p_{out}$.

Energy:

- inlet: $T = T_{in}$,
- pipe wall: $\nabla T = q/k$, where q is the applied wall heat flux,

- outlet: $\nabla T = qD\pi/\left(\dot{m}c_p\right)$,

where D is the inner diameter of the pipe and $\dot{m}$ is the mass flow rate. Thus, for the momentum, there is an inflow boundary at the inlet of the pipe with the prescribed uniform velocity and an outflow boundary at the outlet of the pipe. As for the energy, a uniform temperature is prescribed at the inlet, while the temperature gradient is imposed at the pipe wall which is obtained from the given heat flux. At the outlet, the temperature gradient is imposed which is evaluated theoretically from the analytical solution $\nabla T = qD\pi/\left(\dot{m}c_p\right)$ valid for a constant wall heat flux [39].

Upon the discretization, sets of linear algebraic equations are obtained for each unknown variable, each set of algebraic equations representing a counterpart of the corresponding transport equation:

$$a_P\phi_P + \sum_N a_N\phi_N = b, \tag{17}$$

where the summation is performed over all the neighboring cells N surrounding the cell P of interest and the term b on the right-hand side represents the explicit terms.

The linear systems of algebraic equations are solved to obtain the numerical approximate solutions of the governing transport equations. The system of Equation (17) is solved in an iterative solution procedure, starting from an initial estimate and continually improving the solution in every iteration. The iteration loop is stopped, and the solution is reached when the difference between the solutions in two consecutive iterations is smaller than some small prescribed tolerance.

The proper coupling between the pressure and the velocity in the momentum equation is ensured by using the continuity equation to derive the discrete form of the pressure equation from the discrete form of the momentum equation and iterate via the SIMPLE algorithm for steady-state flows, as implemented in OpenFOAM® [40], which is extended to iterate over the energy equation as well:

1. Set all fields to initial values.
2. Assemble and solve the under-relaxed momentum equation (momentum predictor).
3. Assemble and solve the pressure equation and calculate the conservative volume fluxes, then update the pressure field with under-relaxation and correct the velocity explicitly.
4. Solve the temperature equation using the available volume fluxes, pressure and velocity fields; under-relax the equation implicitly in order to improve convergence.
5. Check convergence for all equations; if the system is not converged, start a new iteration from the step 2.

3. Results

The described computational model is implemented in the foam-extend version of the software OpenFOAM®. The capabilities of the model are first verified by computing a single-phase flow with heat transfer and comparing the computationally obtained results with the existing results in the available literature. After that, the model is used to compute the nanofluid flow and obtain results for various volume fractions of the nanoparticles in the mixture.

For the purpose of evaluation of the heat transfer characteristics, the local Nusselt number and heat transfer coefficient are calculated along the pipe axial axis, as well as the average heat transfer coefficient. The mean temperature of the fluid in a pipe cross-section S is evaluated from the following expression:

$$T_m = \frac{\int_S \rho c_p T \boldsymbol{U} \cdot \mathrm{d}\boldsymbol{S}}{\int_S \rho c_p \boldsymbol{U} \cdot \mathrm{d}\boldsymbol{S}}, \tag{18}$$

The local heat transfer coefficient and the Nusselt number in the pipe cross-section are evaluated from the following expressions:

$$h = \frac{-k\nabla T_{wall}}{T_{wall} - T_m}, \tag{19}$$

$$Nu = \frac{hD}{k}, \tag{20}$$

and the average heat transfer coefficients for the pipe is evaluated from the following expression:

$$\bar{h} = \frac{1}{L}\int_0^L h dx \tag{21}$$

In the above expressions the indices m and $wall$ refer to mean value and value at the pipe wall, respectively, D is the diameter and L is the length of the pipe, and overbar denotes the average value. The local values for T_m, h and Nu are calculated for each column of cells along the axial axis of the pipe. Thus, the number of calculated values in the direction of the axial pipe axis is equal to the number of cells in the axial direction.

For the evaluation of the flow hydrodynamics, the friction factor for laminar flow in the pipe is calculated from the expression:

$$f = \frac{-2\nabla p D}{\rho \bar{v}^2} \tag{22}$$

where $\bar{v}$ is the mean flow velocity in the pipe cross-section, which is equal to the inlet velocity, and ∇p is the local pressure gradient in the same cross-section.

3.1. Verification of the Computational Model

The model is verified for a pure fluid, without nanoparticles, against the known analytical solution [39], the experimental results [38], and the empirical result [41]. In the computational setup, the pipe geometry and fluid (water) properties are the same as in [38], with the Reynolds number of the flow Re = 965, the wall heat flux q = 1000 W/m^2 and the temperature of the fluid at the inlet is T_{in} = 293 K, as listed in Table 1. The fluid in these simulations is Newtonian and the viscous stress tensor is calculated as $\tau = \mu\left[\nabla \boldsymbol{U} + (\nabla \boldsymbol{U})^T\right]$, instead of using the Herschel-Bulkley model. There meshes were used for the verification: 200 × 5, 400 × 10 and 800 × 20 cells in the axial and radial directions, respectively. Figure 3 shows the computationally obtained velocity profile and the friction factor for the laminar flow of water. The results for both, the velocity profile and the friction factor, converge with increasing mesh resolution.

Table 1. Parameters in the computational setup used for the verification.

Pipe Diameter D	Pipe Length L	Inlet Temperature T_{in}	Wall Heat Flux q	Reynolds Number Re
4 mm	5 m	293 K	1000 W/m^2	965

The model predicts very accurately the analytical parabolic velocity profile and the theoretical friction factor, which in this case is f = 64/Re = 0.0663, except at the entrance into the pipe where the boundary layer still develops.

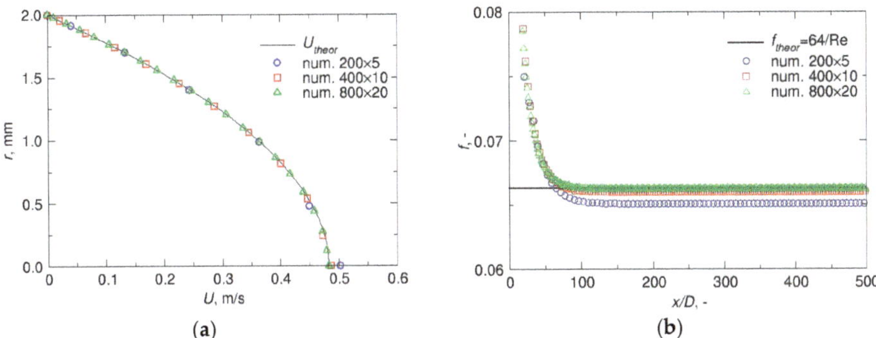

Figure 3. Results for laminar flow of water at $Re = 965$: (**a**) The computed velocity profile evaluated at $x/D = 625$; (**b**) The computed friction factor.

The computed local Nusselt number and the dimensionless radial temperature profile in the pipe cross-section are plotted in Figure 4. The computationally obtained results converge with decreasing the mesh size in both, the dimensionless radial temperature distribution and the Nusselt number. The analytical solution for the Nusselt number $Nu = 4.36$ is also correctly evaluated and the computational results for the Nusselt number are in very good agreement with the experimental results in [38] and also the empirical result from [41].

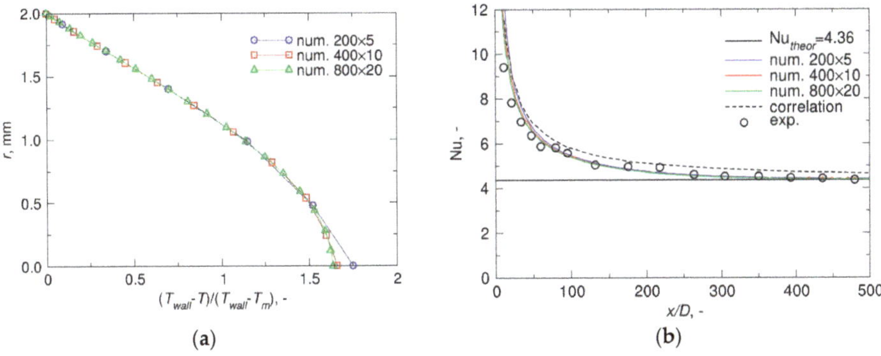

Figure 4. Results for laminar flow of water at $Re = 965$: (**a**) The computed dimensionless temperature profile evaluated at $x/D = 625$; (**b**) The computed Nusselt number.

In order to have an impression on the distributions (fields) of the velocity and the temperature in the pipe, these distributions are shown in Figure 5 for the velocity, and in Figure 6 for the temperature. Due to the large length-to-diameter ratio of the pipe, only results for the parts of the pipe at its inlet and outlet are shown. It can clearly be seen how the velocity and temperature boundary layers develop from the pipe entrance to the pipe outlet. At the pipe outlet the velocity boundary layer is fully developed, while the temperature boundary layer still seems to develop. This is expected, since the hydrodynamic entry length is estimated as $(x_h/D) \approx 0.05Re$ and the thermal entry length is estimated as $(x_t/D) \approx 0.05RePr$ [39], where Pr is the Prandtl number. Thus, for the Prandtl number $Pr > 1$, as is the case here, the velocity boundary layer develops faster than the thermal boundary layer. This is also the reason why the boundary condition for the energy equation at the pipe outlet is given in terms of the gradient of the temperature, which can be calculated analytically and is not a function of the axial coordinate of the pipe [39].

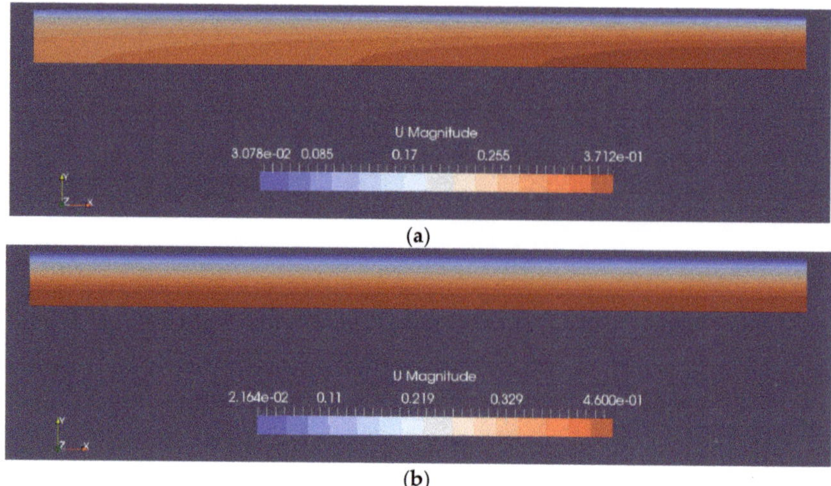

Figure 5. The computed velocity distribution (expressed in m/s) for laminar flow of water at $Re = 965$: (**a**) Inlet part of the pipe; (**b**) Outlet part of the pipe.

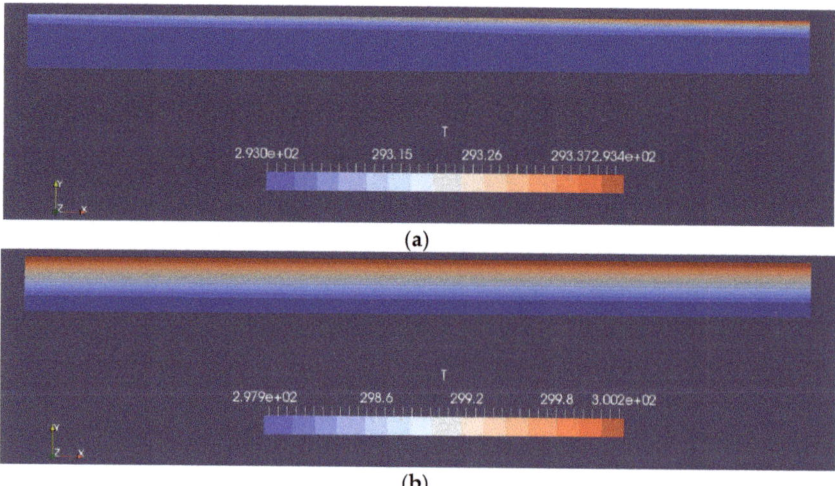

Figure 6. The computed temperature distribution (expressed in K) for laminar flow of water at $Re = 965$: (**a**) Inlet part of the pipe; (**b**) Outlet part of the pipe.

3.2. Results for the EG-Si_3N_4 Nanofluid

After the verification of the computational model in regard to the flow dynamics and heat transfer, further computations of the laminar flow of the EG-Si_3N_4 nanofluid are done. The pipe in the following simulations has a length of 2 m and a diameter of 4 mm, and the applied wall heat flux is equal to 10 kW/m^2. According to the previous results of the verification, and noting that the pipe length is smaller in the simulations with the nanofluid, the numerical mesh consisting of 400×20 cells is taken as sufficient for the simulations. Simulations are performed for various volume flow rates and nanoparticle volume fractions, according to Table 2.

Table 2. Parameters used for the simulation scenarios for ethylene glycol-based silicon nitride ($EG\text{-}Si_3N_4$) nanofluid.

Pipe Diameter D	Pipe Length L	Inlet Temperature T_{in}	Wall Heat Flux q	Volume Flow Rate Q	Nanoparticle Volume Fraction φ_v
4 mm	2 m	293 K	10 kW/m^2	8×10^{-6}–8.4×10^{-5} m^3/s	0–0.035

In Table 2 the volume flow rates of the nanofluid from $Q = 8 \times 10^{-6}$ to 8.4×10^{-5} m^3/s, which were used in the simulations, correspond to the Reynolds numbers in the range from $Re = 200$ to $Re = 2100$ expressed for the base fluid (pure EG). The thermophysical properties for the pure EG as the base fluid and the Si_3N_4 nanoparticles are listed in Tables 3 and 4, respectively.

Table 3. Thermophysical properties of the base fluid (EG).

Density ρ	Specific Heat c_p	Heat Conductivity k
1109.67 kg/m^3	2458.3 J/(kg K)	0.2429 W/(m K)

Table 4. Thermophysical properties of nanoparticles (Si_3N_4).

Density ρ	Specific Heat c_p	Heat Conductivity k
3400 kg/m^3	540 J/(kg K)	16.7 W/(m K)

Figure 7 shows the computationally obtained velocity profile of EG as base fluid and of $EG\text{-}Si_3N_4$ with nanoparticle volume fraction of 3.5% for the volume flow rate $Q = 6 \times 10^{-5}$ m^3/s, which corresponds to $Re = 1500$ for the base fluid. The results reflect the difference in the computed velocity profiles, in particular around the symmetry axis of the pipe, where the profile obtained with the non-Newtonian rheology becomes flattened as expected. It can be concluded that the non-Newtonian shear thinning rheology is correctly predicted by the computational model.

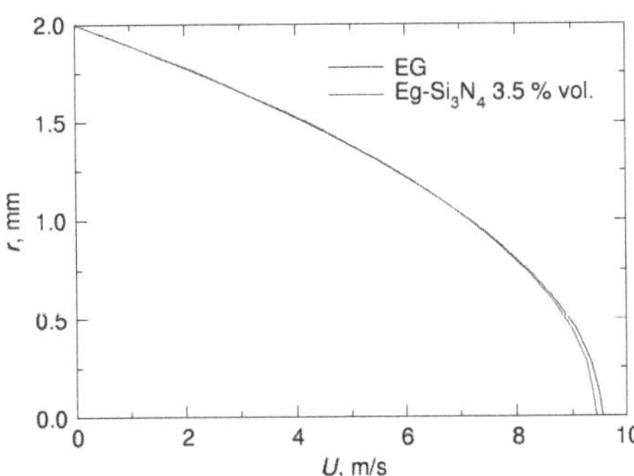

Figure 7. The computed velocity profile for laminar flow of pure EG and of $EG\text{-}Si_3N_4$ nanofluid with the nanoparticle volume fraction of 3.5% at $Q = 6 \times 10^{-5}$ m^3/s, which corresponds to $Re = 1500$ for pure EG.

The computed heat transfer coefficients of EG as base fluid and of $EG\text{-}Si_3N_4$ with different nanoparticle volume fractions are plotted in Figure 8 for the volume flow rates $Q = 8 \times 10^{-6}$ m^3/s, which corresponds to $Re = 200$ for EG, and for $Q = 6 \times 10^{-5}$ m^3/s, which corresponds to $Re = 1500$ for EG. It can be observed that the heat transfer coefficient in the nanofluid is higher than the one in the pure base fluid, and the increase of the volume fraction of the nanoparticles leads to the increase of the heat transfer coefficient.

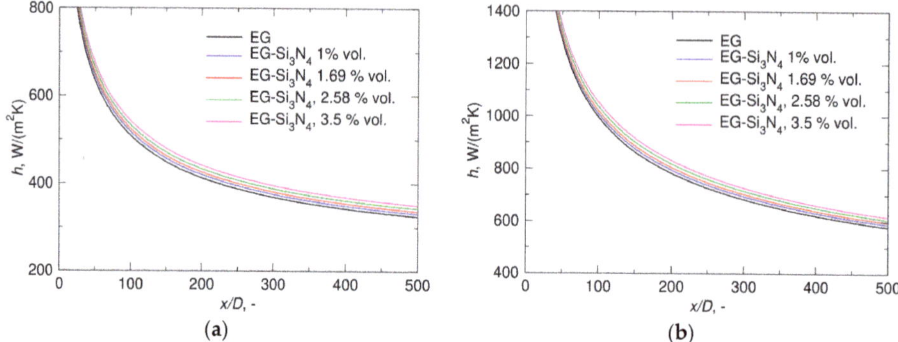

Figure 8. Computational results for laminar flow of EG-Si_3N_4 nanofluid with various nanoparticle volume fractions and the imposed wall heat flux q = 10 kW/m^2: (**a**) Heat transfer coefficient at the volume flow rate of $Q = 8 \times 10^{-6}$ m^3/s, which corresponds to Re = 200 for pure EG; (**b**) Heat transfer coefficient at a volume flow rate of $Q = 6 \times 10^{-5}$ m^3/s, which corresponds to Re = 1500 for pure EG.

Furthermore, simulations were performed with the highest volume fraction of the nanoparticle of ϕ_v = 3.5% in a range of Reynolds numbers between 200 and 2100 (which correspond to the pure EG) and the results for the average heat transfer coefficient in the pipe are shown in Figure 9. As can be seen, the increase in the volume flow rate leads to the increase in the heat transfer coefficient, and the increase of the heat transfer coefficient is slightly more pronounced at higher flow rates.

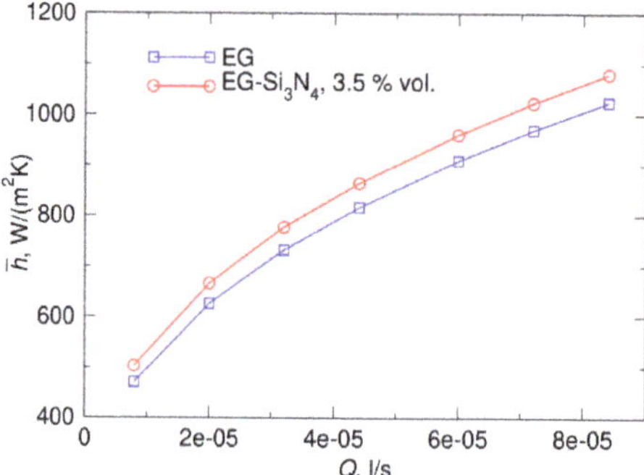

Figure 9. Computational results for the enhancement of the heat transfer coefficient for the EG-Si_3N_4 nanofluid at the nanoparticle volume fraction of 3.5% over a range of nanofluid volume flow rates corresponding to Re between 200 and 2100 for pure EG.

Finally, the computational results for the relative enhancement of the average heat transfer coefficient for the EG-Si_3N_4 nanofluid at the volume fraction of nanoparticles of 3.5% over a range of nanofluid volume flow rates corresponding to Re between 200 and 2100 for pure EG are shown in Figure 10. The relative enhancement of the average heat transfer coefficient is higher at lower flow rates and amounts to 6.5% at the lowest flow rate, while the enhancement is slightly decreasing at higher flow rates. This is attributed to the fact that at higher flow rates the heat transfer coefficient of the pure base fluid is also increased, which slightly damps the relative enhancement.

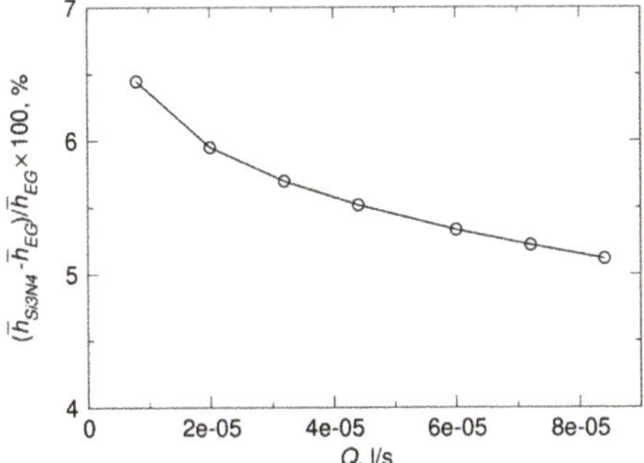

Figure 10. Computational results for the relative enhancement of the heat transfer coefficient for the EG-Si_3N_4 nanofluid at the nanoparticle volume fraction of 3.5% over a range of nanofluid volume flow rates corresponding to *Re* between 200 and 2100 for pure EG.

3.3. Discussion on the Performance Evaluation

The heat transfer enhancement alone should not be used as the only criterion for the evaluation of the heat transfer performance of the nanofluid compared to the pure base fluid. The increase of the viscosity of the nanofluid should also be taken into account, in particular in cases where the change in the viscosity leads to the change in the rheology of the nanofluid. The benefit of using the nanofluid for the heat transfer enhancement in engineering should be estimated bearing in mind the additional pumping power required to compensate for the loss of flow performance due to the increase of the viscosity. Different approaches are proposed in the literature as criteria for the thermal performance evaluation of nanofluids [19,42–44]. They are basically all somehow related to the consideration of the heat transfer performance of the nanofluid compared with the base fluid and relative to the loss of flow power. However, the criteria are proposed by different authors for the specific configuration under consideration (type of the nanofluid, geometry and wall boundary condition) and it seems not to be possible to find a unique criterion straightforwardly. It is therefore advisable to use a criterion which is as simple as possible, while containing information about the relation of the heat transfer increase and the power loss in the nanofluid flow. Since the heat transfer enhancement is directly proportional to the enhancement in thermal conductivity and the loss of the flow power is proportional to the increase in the viscosity, one such criterion for the performance evaluation of the nanofluid is the simple comparison of ratios of thermal conductivities and viscosity of the nanofluid and the base fluid:

$$\frac{k_{nf}}{k_{bf}} > \left(\frac{\mu_{nf}}{\mu_{bf}}\right)^{1/3}, \tag{23}$$

where the indices *nf* and *bf* refer to nanofluid and base fluid, respectively. This criterion can be derived by requiring that the difference between the wall and bulk fluid temperature in the nanofluid flow should be smaller than the one in the flow of the base fluid. It follows from expression (23) that that the specific nanofluid could be considered to be beneficial for use if the nanofluid-to-base fluid thermal conductivity ratio is greater than the third root of the nanofluid-to-base fluid viscosity ratio in the laminar flow regime.

This simple criterion is tested in the present study for various nanoparticle volume fractions and the results are listed in Table 5. The values for the thermal conductivity ratio and the one-third power of the viscosity ratio are relatively close, but the criterion is not satisfied. According to this performance evaluation criterion it would be questionable whether the nanofluid used in the present study would be considered as beneficial for heat transfer applications. In other words, in addition to the gain in the heat transfer performance of the nanofluid, it is important to account for the additional power loss which is in the same order of magnitude but has slightly higher value than the gain in heat transfer.

Table 5. Performance evaluation of the EG-Si_3N_4 nanofluid.

Nanoparticle Volume Fraction φ_v	Thermal Conductivity Ratio k_{nf}/k_{bf}	Viscosity Ratio $(\mu_{nf}/\mu_{bf})^{1/3}$
0.0	1	1
0.0033	1.00947	1.04483
0.0066	1.01894	1.05741
0.01	1.02870	1.07655
0.0134	1.03846	1.11588
0.0169	1.04850	1.13608
0.0258	1.07404	1.24519
0.035	1.100450	1.44602

4. Conclusions

Heat transfer characteristics in laminar flow of the EG-Si_3N_4 nanofluid in a pipe with the imposed constant wall heat flux have been numerically analyzed. The computational model incorporating coupled fluid flow and energy transport has been implemented in the finite-volume numerical approach to enable simulations in the framework of CFD. The nanofluid is modeled as a single-phase mixture, the properties of which are determined based on the volume fraction of nanoparticles. In contrast to most previous studies, the rheology of the nanofluid also depends on the nanoparticle volume fraction and the non-Newtonian shear thinning nanofluid rheology has been taken into account. The model is verified by computing laminar pipe flow coupled with a constant wall heat flux and comparing the numerical results with the available experimental, empirical, and analytical results. The computational model is then used to compute the pipe flow of the EG-Si_3N_4 nanofluid subjected to constant wall heat flux and forced heat convection. Simulated scenarios include various flow rates and nanoparticle volume fractions in the laminar flow regime. The obtained results show that the heat transfer coefficient in the flow of the nanofluid is higher than the one corresponding to the base fluid for the same flow rates, the heat transfer coefficient increases with increasing the volume fraction of nanoparticles and the heat transfer coefficient also increases with increasing the flow rate. The relative enhancement of the average heat transfer coefficient in all cases is higher than 5% and it is about 6.5% at lower flow rates. Although the obtained heat transfer enhancement might appear as lower than expected, this enhancement is still achieved with the EG-Si_3N_4 nanofluid, while on the other hand it is important to note that this nanofluid was recently also found to have very enhanced electrical and optical properties. Thus, it may be considered as suitable for engineering applications where, in addition to the heat transfer, enhanced electrical and optical properties are also important. Furthermore, the present study shows that a relatively simple computational model with the nanofluid being treated as a single-phase mixture offers a convenient and reliable tool for the analysis of nanofluid flow and heat transfer by means of CFD. Finally, it is emphasized that, in spite of the proven capability of the nanofluid to enhance the heat transfer characteristics, it is always very important to objectively asses the performance of the nanofluid in terms of the power loss due to the increase of the viscosity by estimating the nanofluid performance according to an appropriate performance evaluation criterion.

Author Contributions: E.B. and S.B. jointly prepared numerical model, computer simulations, produced computational results, figures, and tables, wrote the draft and finalized the paper. All authors have read and agreed to the published version of the manuscript.

Acknowledgments: This investigation is a contribution to the COST (European Cooperation in Science and Technology) Action CA15119: Overcoming Barriers to Nanofluids Market Uptake (NanoUptake). The authors Edin Berberović and Siniša Bikić acknowledge the EU COST Action CA15119: Overcoming Barriers to Nanofluids Market Uptake for financial support in the participation of the 1st International Conference on Nanofluids (ICNf) and the 2nd European Symposium on Nanofluids (ESNf) participation.

Conflicts of Interest: The authors declare no conflicts of interest.

References

1. Choi, S.U.; Eastman, J.A. *Enhancing Thermal Conductivity of Fluids with Nanoparticles*; Technical report, TRN: 96:001707; Argonne National Laboratory: Argonne, IL, USA, 1995.
2. Bianco, V.; Manca, O.; Nardini, S.; Vafai, K. *Heat Transfer Enhancement with Nanofluids*, 1st ed.; CRC Press: Boca Raton, FL, USA, 2017; pp. 207–287.
3. Huminic, G.; Huminic, A. Application of nanofluids in heat exchangers: A review. *Renew Sustain. Energy Rev.* **2012**, *16*, 5625–5638. [CrossRef]
4. Boukerma, K.; Kadja, M. Convective Heat Transfer of Al2O3 and CuO Nanofluids Using Various Mixtures of Water-Ethylene Glycol as Base Fluids. *Eng. Technol. Appl. Sci. Res.* **2017**, *7*, 1496–1503.
5. Madhesh, D.; Kalaiselvam, S. Experimental study on the heat transfer and flow properties of Ag–ethylene glycol nanofluid as a coolant. *Heat Mass Transf.* **2014**, *50*, 1597–1607. [CrossRef]
6. Wong, K.V.; De Leon, O. Applications of Nanofluids: Current and Future. *Adv. Mech. Eng.* **2010**, *2*, 1–11. [CrossRef]
7. Bhanvase, B.A.; Sarode, M.R.; Putterwar, L.A.; Abdullah, K.A.; Deosarkar, M.P.; Sonawane, S.H. Intensification of convective heat transfer in water/ethylene glycol based nanofluids containing TiO2 nanoparticles. *Chem. Eng. Process.* **2014**, *82*, 123–131. [CrossRef]
8. Sundar, L.S.; Singh, M.K.; Sousa, A.C.M. Enhanced heat transfer and friction factor of MWCNT–Fe3O4/water hybrid nanofluids. *Int. Commun. Heat Mass Transf.* **2014**, *52*, 73–83. [CrossRef]
9. Li, Y.; Zhou, J.; Luo, Z.; Tung, S.; Schneider, E.; Wu, J.; Li, X. Investigation on two abnormal phenomena about thermal conductivity enhancement of BN/EG nanofluids. *Nanoscale Res. Lett.* **2011**, *6*, 1–7. [CrossRef]
10. Ilhan, B.; Kurt, M.; Ertürk, H. Experimental investigation of heat transfer enhancement and viscosity change of hBN nanofluids. *Exp. Therm. Fluid Sci.* **2016**, *77*, 272–283. [CrossRef]
11. Zyła, G.; Fal, J.; Traciak, J.; Gizowska, M.; Perkowski, K. Huge thermal conductivity enhancement in boron nitride–ethylene glycol nanofluids. *Mater. Chem. Phys.* **2016**, *180*, 250–255. [CrossRef]
12. Zyła, G.; Fal, J. Experimental studies on viscosity, thermal and electrical conductivity of aluminum nitride–ethylene glycol (AlN–EG) nanofluids. *Thermochim. Acta* **2016**, *637*, 11–16. [CrossRef]
13. Zyła, G.; Fal, J.; Estellé, P. Thermophysical and dielectric profiles of ethylene glycol based titanium nitride (TiN–EG) nanofluids with various size of particles. *Int. J. Heat Mass Transf.* **2017**, *113*, 1189–1199. [CrossRef]
14. Sarafraz, M.M.; Safaei, M.R.; Tian, Z.; Goodarzi, M.; Filho, E.P.B.; Arjomandi, M. Thermal Assessment of Nano-Particulate Graphene-Water/Ethylene Glycol (WEG 60:40) Nano-Suspension in a Compact Heat Exchanger. *Energies* **2019**, *12*, 1929. [CrossRef]
15. Akbari, M.; Galanis, N.; Behzadmehr, A. Comparative analysis of single and two-phase models for CFD studies of nanofluid heat transfer. *Int. J. Therm. Sci.* **2011**, *50*, 1343–1354. [CrossRef]
16. Ravnik, J.; Škerget, L.; Yeigh, B.W. Nanofluid natural convection around a cylinder by BEM. *WIT Trans. Model. Sim.* **2015**, *61*, 261–271.
17. Ravnik, J.; Škerget, L. Simulation of flow of nanofluids by BEM. *WIT Trans. Model. Sim.* **2010**, *50*, 3–14.
18. Minea, A.A.; Buonomo, B.; Burggraf, J.; Ercole, D.; Karpaiya, K.R.; Di Pasqua, A.; Sekrani, G.; Steffens, J.; Tibaut, J.; Wichmann, N.; et al. NanoRound: A benchmark study on the numerical approach in nanofluids' simulation. *Int. Commun. Heat Mass Transf.* **2019**, *108*, 104292. [CrossRef]
19. Titan, C.P.; Morshed, A.K.M.M.; Khan, J.A. Nanoparticle enhanced ionic liquids (NEILS) as working fluid for the next generation solar collector. *Procedia Eng.* **2013**, *56*, 631–636.
20. Bhattacharjee, A.; Luís, A.; Santos, J.H.; Lopes-da-Silva, J.A.; Freire, M.G.; Carvalhoa, P.J.; Coutinho, J.A.P. Thermophysical properties of sulfonium- and ammonium-based ionic liquids. *Fluid Phase Equilib* **2014**, *381*, 36–45. [CrossRef]
21. Titan, P.C.; Morshed, A.K.M.M.; Fox, E.B.; Visser, A.N.; Bridges, N.J.; Khan, J.A. Thermal performance of ionic liquids for solar thermal applications. *Exp. Therm. Fluid Sci.* **2014**, *59*, 88–95.
22. Titan, P.C.; Morshed, A.K.M.M.; Fox, E.B.; Khan, J.A. Effect of nanoparticle dispersion on thermophysical properties of ionic liquids for its potential application in solar collector. *Procedia Eng.* **2014**, *90*, 643–648.
23. Titan, P.C.; Morshed, A.K.M.M.; Fox, E.B.; Khan, J.A. Thermal performance of Al2O3 Nanoparticle Enhanced Ionic Liquids (NEILs) for Concentrated Solar Power (CSP) applications. *Int. J. Heat Mass Transf.* **2015**, *85*, 585–594.

24. Titan, P.C.; Morshed, A.K.M.M.; Fox, E.B.; Khan, J.A. Enhanced thermophysical properties of NEILs as heat transfer fluids for solar thermal applications. *Appl. Therm. Eng.* **2017**, *110*, 1–9.
25. Chereches, E.I.; Sharma, K.V.; Minea, A.A. A numerical approach in describing ionanofluids behavior in laminar and turbulent flow. *Contin. Mech. Thermodyn.* **2018**, *30*, 657–666. [CrossRef]
26. Minea, A.A.; Sohel Murshed, S.M. A review on development of ionic liquid based nanofluids and their heat transfer behavior. *Renew Sustain Energy Rev.* **2018**, *91*, 584–599. [CrossRef]
27. Mukherjee, S.; Paria, S. Preparation and Stability of Nanofluids-A Review. *IOSR-JMCE* **2013**, *9*, 63–69. [CrossRef]
28. Sharma, B.S.K.; Gupta, S.M. Preparation and evaluation of stable nanofluids for heat transfer application: A review. *Exp. Therm. Fluid Sci* **2016**, *79*, 202–212.
29. Zhao, M.; Lv, W.; Li, Y.; Dai, C.; Zhou, H.; Song, X.; Wu, Y. A Study on Preparation and Stabilizing Mechanism of Hydrophobic Silica Nanofluids. *Materials* **2018**, *11*, 1385. [CrossRef]
30. Alirezaie, A.; Hajmohammad, M.H.; Alipour, A.; Salari, M. Do nanofluids affect the future of heat transfer? A benchmark study on the efficiency of nanofluids. *Energy* **2018**, *157*, 979–989. [CrossRef]
31. Yu, W.; Xie, H.; Li, Y.; Chen, L. Experimental investigation on thermal conductivity and viscosity of aluminum nitride nanofluid. *Particuology* **2011**, *9*, 187–191. [CrossRef]
32. Hussein, A.M.; Noor, M.M.; Kadirgama, K.; Ramasamy, D. Heat transfer enhancement using hybrid nanoparticles in ethylene glycol through a horizontal heated tube. *Int. J. Automot. Mech. Eng.* **2017**, *14*, 4183–4419. [CrossRef]
33. Zyła, G.; Vallejo, J.P.; Lugo, L. Isobaric heat capacity and density of ethylene glycol based nanofluids containing various nitride nanoparticle types: An experimental study. *J. Mol. Liq.* **2018**, *261*, 530–539. [CrossRef]
34. Abbasi, F.M.; Gul, M.; Shehzad, S.A. Hall effects on peristalsis of boron nitride-ethylene glycol nanofluid with temperature dependent thermal conductivity. *Physica E* **2018**, *99*, 275–284. [CrossRef]
35. Zyla, G.; Fal, J.; Bikić, S.; Wanic, M. Ethylene glycol-based silicon nitride nanofluids: An experimental study on their thermophysical, electrical and optical properties. *Physica E* **2018**, *104*, 82–90. [CrossRef]
36. Wikki Consultancy and Software Developement Using Foam-Extend and OpenFOAM®. Available online: http://wikki.gridcore.se/foam-extend (accessed on 24 September 2019).
37. OpenFOAM The Open-Source CFD Toolbox. Available online: https://www.openfoam.com/ (accessed on 24 September 2019).
38. Meyer, J.P.; Everts, M. Single-phase mixed convection of developing and fully developed flow in smooth horizontal circular tubes in the laminar and transitional flow regimes. *Int. J. Heat Mass Transf.* **2018**, *117*, 1251–1273. [CrossRef]
39. Bergman, T.L.; Lavine, A.S.; Incropera, F.P.; Dewitt, D.P. *Fundamentals of Heat and Mass Transf.*, 7th ed.; John Wiley & Sons Inc.: Hoboken, NJ, USA, 2011; pp. 518–558.
40. Jasak, H. Error Analysis and Estimation for the Finite Volume Method with Applications to Fluid Flows. Ph.D. Thesis, Imperial College of Science, Technology and Medicine, London, UK, June 1996.
41. Shah, R.K.; London, A.L. *Laminar Flow Forced Convection in Ducts*, 1st ed.; Academic Press: New York, NY, USA, 1978; pp. 78–138.
42. Prasher, R.; Song, D.; Wang, J.; Phelan, P. Measurements of nanofluid viscosity and its implications for thermal applications. *Appl. Phys. Lett.* **2006**, *89*, 133108. [CrossRef]
43. Wu, Z.; Feng, Z.; Sundén, B.; Wadsö, L. A comparative study on thermal conductivity and rheology properties of alumina and multi-walled carbon nanotube nanofluids. *Front. Heat Mass Transf.* **2014**, *5*, 1–10.
44. Sekrani, G.; Poncet, S.; Proulx, P. Modelling of convective turbulent heat transfer of water-based Al_2O_3 nanofluids in a uniformly heated pipe. *Chem. Eng. Sci.* **2018**, *176*, 205–219. [CrossRef]

Review

Towards the Correct Measurement of Thermal Conductivity of Ionic Melts and Nanofluids †

Carlos A. Nieto de Castro * and Maria José V. Lourenço

Centro de Química Estrutural, Faculdade de Ciências, Universidade de Lisboa, Campo Grande, 1749-016 Lisbon, Portugal; mjlourenco@ciencias.ulisboa.pt

* Correspondence: cacastro@ciencias.ulisboa.pt

† Presented at the 1st International Conference on Nanofluids, Castelló, Spain, 26–28 June 2019.

Received: 19 November 2019; Accepted: 19 December 2019; Published: 24 December 2019

Abstract: Thermophysical properties of engineering fluids have proven in the past to be essential for the design of physical and chemical processing and reaction equipment in the chemical, metallurgical, and allied industries, as they influence directly the design parameters and performance of plant units in the of, for example, heat exchangers, distillation columns, phase separation, and reactors. In the energy field, the search for the optimization of existing and alternative fuels, either using neutral or ionic fluids, is an actual research and application topic, both for new applications and the sustainable development of old technologies. One of the most important drawbacks in the industrial use of thermophysical property data is the common discrepancies in available data, measured with different methods, different samples, and questionable quality assessment. Measuring accurately the thermal conductivity of fluids has been a very successful task since the late 1970s due to the efforts of several schools in Europe, Japan, and the United States. However, the application of the most accurate techniques to several systems with technological importance, like ionic liquids, nanofluids, and molten salts, has not been made in the last ten years in a correct fashion, generating highly inaccurate data, which do not reflect the real physical situation. It is the purpose of this paper to review critically the best available techniques for the measurement of thermal conductivity of fluids, with special emphasis on transient methods and their application to ionic liquids, nanofluids, and molten salts.

Keywords: energy applications; heat transfer; thermal conductivity; measuring methods; nanofluids; molten salts; ionic liquids

---

1. Introduction

Thermal conductivity has been proven to be one of the most difficult properties of materials to be measured with high accuracy. This fact is due to the different molecular mechanisms of heat transfer in solids, liquids, and gases, with neutral or ionic media, and to the difficulties involved in the isolation of pure conduction from other mechanisms of heat transfer, like convection and radiation, a fact that arises from the contradictory requirements of imposing a temperature gradient on the fluid while preventing its motion. This is particularly important in steady state measurements, where the temperature in any point of the measured sample is time independent, and where the heat flux necessary to keep a temperature difference between two surfaces is measured. For isotropic fluids, the thermal conductivity is defined by the Fourier's law, depends on the thermodynamic state of the fluid prior to the perturbation, and must be related with a reference state, not necessarily equal to the initial one [1].

$$q = -\lambda \nabla T \quad (1)$$

where q is the instantaneous flux of heat, the response of the medium to the instantaneous temperature gradient ∇T, and λ the thermal conductivity of the medium. As it is impossible to measure local fluxes and local gradients, the Fourier equation cannot be used directly, and the energy equation must be adapted to a given geometry. The equation of energy conservation in the system is the basis for the formulation of the working equation of any method of measurement of thermal conductivity. The equation of change for non-isothermal systems can be found on the excellent book by Bird, Stuart, and Lightfoot [2], which is written for unit mass as:

$$\rho\frac{\mathrm{D}U}{\mathrm{D}t} = -(\nabla\cdot\boldsymbol{q}) - P(\nabla\cdot\boldsymbol{v}) - (\boldsymbol{\tau}:\nabla\boldsymbol{v}) \tag{2}$$

where ρ is the fluid density, U its internal energy, t the time, P the hydrostatic pressure, v the hydrodynamic velocity of the fluid, and τ the stress tensor. This equation does not include nuclear, radiative, electromagnetic, or chemical terms for energy. In addition, it is applicable to Newtonian fluids; otherwise the last term of Equation (2) must be reformulated for viscoelastic fluids. The notation $\mathrm{D}/\mathrm{D}t$ represents the substantial derivative, meaning that the time rate of change is reported as one move with the fluid (The substantial derivative is given by $\frac{\mathrm{D}}{\mathrm{D}t} = \frac{\partial}{\partial t} + \boldsymbol{v}\cdot\nabla$). From a point of view of experiment, the temperature time evolution is more convenient to obtain. After some manipulation, using the definition of enthalpy and heat capacity at constant pressure, C_P, we can obtain the equation of change for temperature:

$$\rho C_P\frac{\mathrm{D}T}{\mathrm{D}t} = -(\nabla\cdot\boldsymbol{q}) - (\boldsymbol{\tau}:\nabla\boldsymbol{v}) - \left(\frac{\partial\ln\rho}{\partial\ln T}\right)_P \tag{3}$$

If the sample is not moving (solid of stationary fluid, no convection), the properties of the sample do not vary with temperature (small temperature gradients), and using the Fourier law (Equation (1)), this last equation cane be transformed to:

$$\rho C_P\frac{\partial T}{\partial t} = \nabla^2 T \tag{4}$$

where ∇^2 is the symbol for the Laplacian operator. It is important to recognize that transport of energy by radiation is always present, and must be corrected for each measuring technique, especially if measurements are performed at high temperatures. Then, terms corresponding to the radiative energy absorption and emission rates must be added. Equation (4) is the basis of all experimental methods for the measurement of thermal conductivity.

Between 1950 and 1990, a variety of experimental methods have been developed, for gaseous, liquid, supercritical or solid phases, over wide range of thermodynamic states. Until the 1970s much of the experimental work on the thermophysical properties of fluids was devoted to the development of methods for the measurement of the properties of simple fluids under moderate conditions. By the end of the 1970s a few methods emerged that had both a rigorous mathematical description of the experimental method and technical innovation to render measurements precise enough to rigorously test theories of fluids for both gas and liquid phases. It was also possible with the development of electronics and fast data acquisition systems, to establish highly accurate instrumentation and methods for fluids (gases and liquids) and solids for wide temperature and pressure ranges and the establishment of standard reference data. The reader can obtain a lot of information in references [3–6].

These methods are based on the simplified energy equation and can be classified in two main categories:

- Unsteady state or transient methods, in which the full Equation (4) is used and the principal measurement is the temporal history of the fluid temperature (transient hot-wire, transient hot strip, the interferometric technique adapted to states near the critical point etc.);

- Steady state techniques, for which $\frac{\partial T}{\partial t} = 0$ and the equation reduces to $\nabla^2 T = 0$, which can be integrated for a given geometry (parallel plates, concentric cylinders, etc.).

However, it is difficult and tedious to make good measurements of thermophysical properties. Nowadays it is also neither fashionable nor economically attractive for industries and funding agencies, and the need for accurate data on the thermophysical properties of fluids has been replaced to a certain extent by alternative methods of properties calculation, mostly based on computer simulation or prediction/estimation methodologies (cheaper, but risky), and by the apparent contradiction between accurate (low uncertainty) or fit for purpose. This last approach supported the development of many commercial types of equipment, which were developed with the main objective of being user friendly and fast.

Reasons for this can be summarized in the following items:

- Change of needs—From laboratory work to in-situ measurements,
- Change of paradigm—Fit for purpose instead of best uncertainty,
- Change of financing priorities from state funding agencies—priority to industry driven/sponsored research,
- Decrease for users in industry of the added value of property data of good quality,
- The use and misuse of commercial equipment's, with methods of measurement not adequate to the object systems.

Current practice shows that, despite being based on established methods, construction, and type of fluid/solid to be applied to, several inaccuracies exist in the data obtained and sometimes to completely erroneous results. This is the case of their application in the fluids area to new and more complex systems like humid air, nanofluids, ionic melts (molten salts and ionic liquids) and nanofluids.

Figure 1 shows the situation for the thermal conductivity of molten alkali nitrates in 2016. Data was selected accordingly to our selection of "best data". Details of the authorship and references can be found in the paper by Nunes et al., 2016 [7]. It shows important scattering of data between different authors for de same salt/mixture, much greater than the claimed mutual uncertainty of data. An extraordinary example is the data presented in Janz NIST database (1979) [8] and the paper by Nagasaka et al., 1991 for potassium nitrate (KNO_3), where positive or negative slopes can be observed [9].

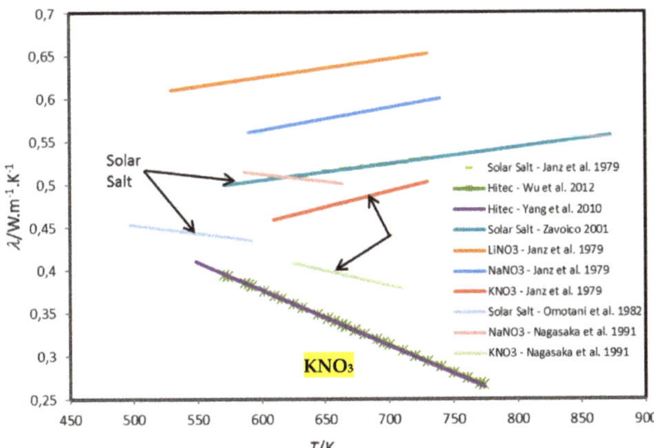

Figure 1. Thermal conductivity of molten alkali nitrates and mixtures. Solar salt is a eutectic binary mixture ($NaNO_3$:KNO_3 60/40 wt%), while HITEC® is a eutectic ternary mixture ($NaNO_2$: KNO_3: $NaNO_3$ 7/53/40 wt%). Adapted from [7].

It is the purpose of this paper to review critically the best available techniques for the measurement of thermal conductivity of fluids, with special emphasis on transient methods and their application to ionic liquids, nanofluids, and molten salts. In addition, several remarks on how theory can help to access the quality of data obtained.

2. Methods of Measuring Thermal Conductivity of Ionic Melts and Nanofluids

It is not the purpose of this paper to describe in detail all the methods developed so far for the measurement of the thermal conductivity, namely for the temperature range 300–1500 K, at ambient pressure, the main zone where ionic liquids, molten salts and nanofluids are important and have several industrial applications. Several of these methods can be applied also to lower temperature melts/systems, and to solids. A complete discussion of the situation can be seen in the paper by Mardolcar and Nieto de Castro, 1992 [5]. From classical methods (transient hot-wire, concentric cylinders or parallel plates) to non-classical methods (laser flash, forced Rayleigh scattering, photon correlation spectroscopy, photothermal radiometry, transient thermoreflectance, photothermal detection, wave-front-shearing interferometry, heat flow, AC calorimetry, photoacoustic and radiation heat exchange), all are there described, instrumentally and mathematical modelling of the different physical situations, identifying all the advantages and disadvantages of each method. To these, we must add the transient hot strip method, the transient plane source or hot-disk and the 3ω method, all developed after 1992.

The methods are classified as primary and secondary, as explained in the reference [10] (The definition of primary method was approved by CCQM—*Comité Consultatif pour la Quantité de Matière*, BIPM, in 1995—"a primary method of measurement is a method having the highest metrological qualities, the operation of which can be completely described and understood, for which a complete uncertainty statement can be written down in terms of SI units, the results of which are therefore accepted without reference to a standard of the quantity being measured."). The values quoted for the attainable uncertainty correspond to the "state-of-art" of the method and cannot be compared with commercial instruments, which as mentioned above, sacrifice high accuracy for ease of operation and cost of equipment. Transient methods are the only ones that can compete for the statute of primary methods, in the sense of CIPM[2]. These methods include the transient hot-wire and the transient hot strip, the latter still requiring the solution of 3D heat transfer equation for the geometries involved.

Table 1 displays selected methods for the measurement of thermal conductivity (thermal diffusivity) and its applicability to ionic melts and nanofluids, after a careful selection by the authors of the most likely to have success in the measurements. This methodology was applied before by the authors in two different papers, one dedicated to low temperature ionic liquids and molten salts [10], and more recently to nanofluids [11], and it is improved here for comparison. Reference must also be made to four methods for the determination of thermal diffusivity, $\alpha = \frac{\lambda}{\rho C_P}$, which are not direct measurements of thermal conductivity, needing accurate values of density (usually available) and heat capacity (more difficult to obtain), the laser flash [12,13], the photon-correlation spectroscopy [14], the forced Rayleigh scattering method [15], and the transient grating technique [16]. The first one has potential for high temperature melts and nanofluids, the second one is a very versatile an amenable to improvement method, the third one has been applied to molten salts and the fourth one contains the first reported measurements for the thermal conductivity of LTILs. Some discussion of the methods applicable to nanofluids can be found in the work of Bobbo and Fidele [17].

Table 1. Selected methods for the measurement of thermal conductivity (thermal diffusivity) and its applicability to ionic melts and nanofluids.

Method	Type	Attainable Uncertainty [1]	Applicability to Ionic Melts			Applicability to Nanofluids		
			Yes	No	Remarks	Yes	No	Remarks
Transient hot-wire for conducting liquids [2]	Primary	1%	✓		The best method to be applied with classical designs	✓		The best method to be applied with classical designs
Transient hot-wire (bare wires)	Primary	1%		✓		✓		Non-conducting base fluids and/or nanoparticles
Transient hot strip for conducting liquids (insulated strip)	Primary/Secondary	2–3%	✓		Can be considered primary if the 3D heat transfer equation is solved for the geometries involved	✓		Can be considered primary if the 3D heat transfer equation is solved for the geometries involved
Transient Plane Source or Hot-Disk	Secondary	3–5%	✓		Limited in temperature range	✓		Limited in temperature range
Steady-state Parallel Plates (Guarded Hot-Plates)	Secondary	2–3%	✓			✓		Small spacing between plates might induce phase separation
Steady-state Concentric Cylinders	Secondary	2–3%	✓		Small gaps. Guarded plates	✓		Small gaps must be used.
Laser Flash	Secondary	3–5%	✓		Difficulties with originated waves in the liquid surface (Marangoni effects)	✓		Use low power lasers to avoid NP breakage and structure deployment
Forced Rayleigh Scattering	Secondary/Relative	3–5%	✓		Dyes necessary to enhance the signal compatible with molten salts or IL's	✓		Dyes necessary to enhance the signal compatible with nanofluids.
Photon Correlation Spectroscopy	Secondary/Relative	2–3%	✓		Optically transparent fluids. Not explored for measurements above 473 K.	✓		Optically transparent fluids.
Transient Grating	Secondary/Relative	3–5%		✓	Needs a big improvement in the calibration. Not recommended for high quality work		✓	Needs a big improvement in the calibration. Not recommended for high quality work
3ω Method	Secondary	5%	✓		Needs improvement of the theory. However, alternating current, destroying the direction of the polarizing current, avoids current polarization	✓		Needs improvement of the theory. However, alternating current, destroying the direction of the polarizing current, avoids nanoparticle deposition and electrically insulation problems.

[1] Uncertainty is defined using ISO criteria, using a coverage factor (k = 2), that correspond to a 95% confidence level. [2] Both the insulated probe and the polarization technique can be used.

From all these methods, only the transient hot-wire can be today considered as a primary method. In fact [18]:

- The transient hot wire technique was identified as the best technique for obtaining standard reference data (certification of reference materials).
- It is an absolute technique, with a working equation and a complete set of corrections reflecting the departure from the ideal model, where the principal variables are measured with a high degree of accuracy. It is accepted by the scientific community as a primary method, the top of the traceability chain for this physical quantity.
- The liquids proposed by IUPAC (toluene, benzene, and water) as primary standards for the measurement of thermal conductivity were measured with this technique with an accuracy of 1% or better.
- It has been extended, from its original version, to its application at high temperatures (correction for radiation effects), for conducting liquids and melts.

Although several texts in the past have described the complete theory of the method and its application to gases and liquids, we would like to make a short description because many commercial applications, although mentioning that their method is based on the hot-wire principles, do not pay attention to many deviations from the ideal mathematical model.

The theory of the transient hot wire is well known, and a complete uncertainty analysis has been previously presented [19–21]. The ideal mathematical model of an infinite vertical line heat source immersed in an infinite isotropic medium with properties independent of the temperature, and in thermodynamic equilibrium T_0 at $t = 0$, and when a stepwise heat flux q per unit length is applied to the wire, generates a pure conductive heat flux that will transfer from the wire to the immersed medium, obtained from the solution of Equation (4), that will increase as a function of time the temperature at the surface of the wire, decaying in the media, as a function of time and distance to the centre of the wire, r. This temperature rise can be defined as:

$$\Delta T(r,t) = T(r,t) - T_0 \tag{5}$$

Equation (4) can be solved subjected to boundary conditions:

$$\begin{gathered} \Delta T(r,t) = 0 \ \text{ for } t \leq 0, \ \text{ any } r \\ \lim_{r\to 0} \Delta T(r,t) = 0 \text{ for } t > 0 \\ \lim_{r\to 0} r\frac{\partial T}{\partial r} = -\frac{q}{2\pi\lambda} = \text{constant for } t \geq 0 \end{gathered} \tag{6}$$

where the additional condition that the thermal diffusivity $\alpha = \frac{\lambda}{\varrho C_p}$ is constant [22]. The result for the ideal temperature rise is:

$$\Delta T_{\text{id}}(r,t) = T(r,t) - T_0 = -\frac{q}{4\pi\lambda}E_1\left(\frac{r^2}{4\alpha t}\right) \tag{7}$$

where $E_1(x)$ is the exponential integral. This integral can be expanded by:

$$E_1\left(\frac{r^2}{4\alpha t}\right) = \int_{\frac{r^2}{4\alpha t}}^{0} \frac{e^{-y}}{y}dy = -\gamma - \ln\left(\frac{r^2}{4\alpha t}\right) + \frac{r^2}{4\alpha t} + O\left[\left(\frac{r^2}{4\alpha t}\right)^2\right] \tag{8}$$

where $\gamma = 0.5772157\ldots$ being the Euler's constant. If the line source is replaced by a cylindrical wire of radius r_i (as small as possible) and assuming that the temperature of the fluid in the ideal model is equal to the temperature at the surface of this wire, at $r = r_i$. Then, for small values of $\frac{r^2}{4\alpha t}$, we can obtain:

$$\Delta T_{\text{id}}(r_i,t) = T(r_i,t) - T_0 = \frac{q}{4\pi\lambda}\ln\left(\frac{4\alpha t}{r_i^2 C}\right) + \frac{r_i^2}{4\alpha t} + \ldots \tag{9}$$

where $C = \exp(\gamma)$. The radius must be as small as possible and therefore if the term $\frac{r_i^2}{4\alpha t} < 0.01\%$ of ΔT_{id}, then the temperature increases at the surface of the hot-wire is given by:

$$\Delta T_{\mathrm{id}}(r_i, t) = T(r_i, t) - T_0 = \frac{q}{4\pi\lambda} \ln\left(\frac{4\alpha t}{r_i^2 C}\right) \tag{10}$$

Equation (10) is the fundamental working equation of the transient-hot wire technique, if the conditions used for its derivation are realized experimentally. If so, the thermal conductivity can be obtained from the slope of the straight line ΔT_{id}versus logarithm of time and referred to the temperature $T(r_i, t)$, if measured experimentally.

However, the real model of the transient hot wire constructed in any instrument has departures from the ideal mathematical model, namely the finite diameter of the wire, its finite physical properties, its finite length, and its supports. Using the transient hot wire method, the uncertainty of the experimental determined thermal conductivity derives from uncertainty associated to other modes of heat transfer such as convection and radiation, and from the uncertainty associated with random and systematic errors in the measurement of input quantities [20]. Consequently, the experimental measurements of the temperature rise of the wire, ΔT_{w}, depart from the ideal temperature rise ΔT_{id} predicted by Equation (10) and a set of small corrections δT_i must be added to the actual temperature rise, given by:

$$\Delta T_{\mathrm{id}}(r_i, t) = \Delta T_{\mathrm{w}}(r_i, t) + \sum_i \delta T_i \tag{11}$$

Most of these corrections were devised by Healy et al. [23]. Some effects can be minimized by proper design of the instrument, but others must be accounted for. The various corrections δT_i are summarized elsewhere [19,24], the contribution to the global uncertainty of the thermal conductivity of these corrections never amount to more than 0.1% of thermal conductivity [21], when a proper design is made. However, Healy et al. [23] also noted that the thermal conductivity must be assigned, not to T_0 but to a reference thermodynamic state (T_{ref}, ρ_{ref}) defined as:

$$\begin{aligned} T_{\mathrm{ref}} &= T_0 + \sum_i \delta T_i^* \\ \varrho_{\mathrm{ref}} &= \varrho(T_{\mathrm{ref}}, P_0) \end{aligned} \tag{12}$$

These corrections δT_i^* are essentially two. δT_1^* refers to the fact that the bulk fluid properties are temperature dependent. This fact causes one correction, that is time dependent and is better used in the correction to the temperature rise (δT_7) and another part, constant in time, δT_1^* given by:

$$\delta T_1^* = \frac{1}{2}\left[\Delta T(t_i) + \Delta T(t_f)\right] \tag{13}$$

Here t_{i} and t_{f} are the initial and final times of the measurement interval used in the regression line. The second correction $\delta T_2^*$is to be utilized if the metal wire is coated, with a metal oxide or a polymer to insulate electrically the wire from an electrically conducting media (like an ionic liquid, an electrolyte solution or a molten salt), and expressions were proposed in first place by Nagasaka and Nagashima in 1981 [25]. In this case, the value of $\Delta T(t_2)$ in Equation (13) must be calculated at the surface of the coat [25]. A systematic analysis of the departures us presented in references [19,24], and here in Table 2 with the necessary adaptations.

Table 2. Departures from ideal model of transient hot-wire technique.

Correction	Physical Source	Expression	Reference	Recommendations
δT_1	Wire physical properties	$\delta T_1 = \frac{q}{4\pi\lambda}\left\{\frac{r_i^2[(\rho C_P)_w-(\rho C_P)]}{2\lambda t}\ln\left(\frac{4\alpha t}{r_i^2 C}\right)-\frac{r_i^2}{2\alpha t}+\frac{r_i^2}{4\alpha_w t}-\frac{\lambda}{2\lambda_w}\right\}$	[22,23]	Apply always correction.
δT_2	Outer boundary. Cell physical dimensions (b is cell wall diameter)	$\delta T_2 = \frac{q}{4\pi\lambda}\left\{\ln\left(\frac{4\alpha t}{b^2 C}\right)+\sum_{v=1}^{\infty} e^{-g_v \alpha t/b^2}[\pi Y(g_v)]^2\right\}$	[23,26]	Apply always correction. Make design of cell to minimize it to 0.1% of ΔT_{id}.
δT_3	Compression work (L is the wire length and V is the cell volume)	$\delta T_3 = -\left[\frac{qLRt}{\rho C_P C_V V}\left(1-e^{-b^2/4\alpha t}\right)+\frac{qR}{4C_V\lambda}E_1\left(\frac{b^2}{4\alpha t}\right)\right]$	[23,24]	Negligible in liquid phases. For gases limits the operational zone of measurements to $\rho > 40$ kg·m^{-3}.
δT_4	Radial convection and viscous dissipation. Symbol w means wire properties. (α_P is the isobaric thermal expansion coefficient of fluid and g the acceleration of gravity)	For gases and liquids ($\Delta\rho < \rho_0$) $\delta T_4 = -\left(\frac{q}{4\pi\lambda}\right)^2 \alpha_P\left[\ln 4 + \frac{\mathbf{Pr} g^2 t^2 \alpha_P}{2C_P}\right]$ For liquids, expressions also graphical form, recording the vertical penetration depth of moving fluid front. $Z_p^* = Z_p \frac{\lambda}{g\alpha_P q t^2} = Z_p^*(\mathbf{Pr}, t^*, r^*, \omega)$ $\mathbf{Pr} = \frac{\eta C_P}{\lambda}$; $t^* = \frac{\alpha t}{r_i^2}$; $r^* = \frac{r}{r_i}$; $\omega = \frac{2\rho C_P}{(\rho C_P)_w}$	[23,24,27–30]	Avoid onset of convection in the measurements. Design measurement time to $t < 5$ s to render negligible in liquid phases. Wire(s) must be vertical. Avoid short wires.
δT_5	Radiation (s is the Stefan-Boltzmann constant, K is the a mean extinction coefficient for radiation and n the refractive index of the liquid)	Transparent fluid $\delta T_5 = \frac{8\pi r_i \sigma T_0^3 \Delta T_{id}^2}{q}$ Absorbing fluid $\delta T_5 = -\frac{qB}{4\pi\lambda}\left[\frac{r_i^2}{4\alpha}\ln\left(\frac{4\alpha t}{r_i^2 C}\right)+\frac{r_i^2}{4\alpha}-t\right]$ *with* $B \simeq \frac{16Kn^2\sigma T_0^3}{\rho C_P}$	[23,31,32]	Always apply correction, dependent if fluid is transparent or absorbs radiation. K is difficult to obtain experimentally, but usually $Br_i^2/4\alpha < 10^{-5}$ [30,32]. Introduces curvature in straight line and it is mostly important at high temperatures (molten salts).
δT_6	Temperature jump	$\delta T_6 \simeq \frac{q\Lambda}{2\pi\lambda r_i}\frac{\Delta T_{id}}{T_0}$	[23,29]	Important for gases. Work outside density range where it is not negligible (mean free path of the gas molecules Λ approaching wire diameter (microns). For nanofluids might play a role, but here with nanoparticles dimension.

Table 2. *Cont.*

Correction	Physical Source	Expression	Reference	Recommendations
δT_7	Effect of variable fluid properties	$\delta T_7 = -\left[\frac{q}{4\pi\lambda(T_0)}\right]^2(\chi-\gamma)\ln 4$ $\lambda(T,\rho) = \lambda(T_0)(1+\chi\Delta T) \quad \rho C_P(T,\rho) = \rho C_P(T_0)(1+\gamma\Delta T)$	[24,33]	Always apply correction. Valid for situations where fluid density during a measurement can be considered constant, as in the case of systems far from critical region.
δT_8	Effect of variable fluid properties (fluids critical region)	$\delta T_8 = -(\chi'' - \gamma'')\left(\frac{q}{4\pi\lambda(T_0)}\right)^3 \ln 4 - \frac{\chi''}{6}\Delta T^3$ $\chi'' = \phi\rho_0\alpha_P^2;\ r'' = \varepsilon\rho_0\alpha_P^2;\ \varepsilon = \frac{1}{\rho C_P(T_0)}\left[\frac{\partial(\rho C_P)}{\partial\rho}\right]_T$	[30]	Always apply correction. For measurements in the critical region, the properties are extremely strong functions of temperature.
δT_9	Truncation error resulting from the expansion of exponential integral, Equation (8)	$\delta T_9 = \Delta T_{id}\frac{r_i^2}{\left[4\alpha t\ln\left(\frac{4\alpha t}{r_i^2 C}\right)\right]}$	[30]	Rendered negligible by design. It decreases with time.
δT_{10}	Effect of coating to insulate wires for electrically conducting liquids. Subscript 2 means coating properties. r_0 is coating radius	$\delta T_9 = -\frac{B}{t}\Delta T_{id} + \frac{q}{4\pi\lambda}\left\{\ln\left(\frac{r_0}{r_i}\right)^2\left(1-\frac{\lambda}{\lambda_2}\right) - \frac{\lambda}{2\lambda_w} - \frac{1}{t}\left[D - B\ln\frac{4\lambda}{r_i^2 C}\right]\right\}$ $B = \frac{1}{2\lambda}\left[r_i^2\left(\frac{\lambda_2}{\alpha_2}-\frac{\lambda_w}{\alpha_w}\right) + r_0^2\left(\frac{\lambda}{\alpha}-\frac{\lambda_2}{\alpha_2}\right)\right]$ $D = \frac{r_i^2}{8}\left[\left(\frac{\lambda-\lambda_2}{\lambda_w}\right)\left(\frac{1}{\alpha_w}+\frac{1}{\alpha_2}\right) + \frac{4}{\alpha_2} - \frac{2}{\alpha_w}\right] + \frac{r_0^2}{2}\left(\frac{1}{\alpha}+\frac{1}{\alpha_2}\right)$ $+\frac{r_i^2}{\lambda_2}\left(\frac{\lambda_2}{\alpha_2}-\frac{\lambda_w}{\alpha_w}\right)\ln\left(\frac{r_0}{r_i}\right)$ $+\frac{1}{2\lambda}\left[r_i^2\left(\frac{\lambda_2}{\alpha_2}-\frac{\lambda_w}{\alpha_w}\right) + r_0^2\left(\frac{\lambda}{\alpha}-\frac{\lambda_2}{\alpha_2}\right)\right]\ln\frac{4\alpha}{r_0^2 C}$	[25,34]	Always apply correction.
δT_1^*	Effect of variable fluid properties	$\delta T_1^* = \frac{1}{2}\left[\Delta T(t_i) + \Delta T(t_2)\right]$	[4,32,33]	Always apply correction. When the distribution of the measured temperature rises is not uniform, a more detailed expression is necessary [32].
δT_2^*	Effect of coating to insulate wires for electrically conducting liquids	$\delta T_2^* = \frac{q}{8\pi\lambda}\left\{2\ln\left(\frac{r_0}{r_i}\right)^2 - r_i^2\left[\left(\frac{1}{t_1}+\frac{1}{t_2}\right)\right]\left[\left(\frac{1}{\alpha_w}+\frac{1}{\alpha_2}\right) + \left(\frac{r_0}{r_i}\right)^2\left(\frac{1}{\alpha}-\frac{1}{\alpha_2}\right)\right]\right.$ $+\frac{2}{\lambda_2}\left(\frac{\lambda_2}{\alpha_2}-\frac{\lambda_w}{\alpha_w}\right)\ln\frac{r_0}{r_i} - \frac{2}{\lambda}\left[\left(\frac{\lambda_2}{\alpha_2}-\frac{\lambda_w}{\alpha_w}\right)\right.$ $\left.+\left(\frac{r_0}{r_i}\right)^2\left(\frac{\lambda}{\alpha}-\frac{\lambda_2}{\alpha_2}\right)\right]\ln\frac{r_0}{r_i}$ $+\frac{2}{\lambda}\left[\left(\frac{\lambda_2}{\alpha_2}-\frac{\lambda_w}{\alpha_w}\right) + \left(\frac{r_0}{r_i}\right)^2\left(\frac{\lambda}{\alpha}-\frac{\lambda_2}{\alpha_2}\right)\right]\left[\frac{1}{t_1}\ln\frac{4\alpha t_1}{r_i^2 C}\right.$ $\left.\left.+\frac{1}{t_2}\ln\frac{4\alpha t_2}{r_i^2 C}\right]\right\}$	[25,34]	Always apply correction. Remember to correct the expression for $\Delta T(t_2)$ in δT_1^*, which has to be evaluated at the surface of the coat ($r = r_0$).

Several comments must be made. Some of the corrections have to be applied to have accurate measurements (δT_1, δT_2, δT_5, δT_7, δT_8, δT_{10}, δT_1^*, δT_2^*), and some others must be rendered negligible by a careful instrument design (δT_3, δT_6, δT_9). Once this approach is made, it is possible to obtain straight lines with very good statistics, like those shown in Figure 2. This data was obtained with an instrument specially developed in Lisbon for corrosive environments, humid air, with an expanded uncertainty ($k = 2$) of 1.7%. The hot-wire cells are presented in Figure 3 and details of it can be find in reference [35].

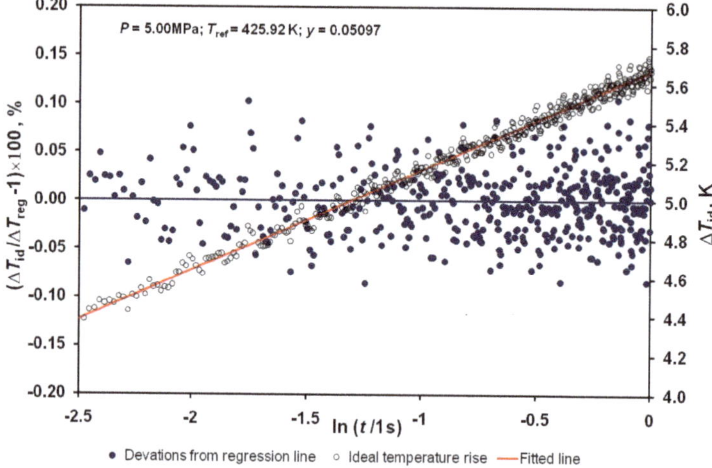

Figure 2. Deviations between the ideal temperature rises and those obtained from the linear regression line—Scattering plot, for a test with humid air, for $T_{ref} = 425.92$ K, P = 5.00 MPa and $y_{H2O} = 0.05097$. Adapted from [35].

A more recent, alternative approach to the approximate analytic solution described above is to solve the full Fourier heat transfer equations in the wire and the test material, using numerical methods based upon the finite element method (FEM). This procedure was developed and validated with water by Antoniadis et al. [36], using a two-tantalum hot wires system. Agreement between the two procedures was within 0.2% at 297.5 K and atmospheric pressure. Both the FEM method and the approximate analytic solution herein described have uncertainties of 0.5%. Applications to nanofluids of TiO_2 and MWCNTs showed that, when the measurements are judiciously performed, with a validated "state of art" instrument, the results can even prove which earlier results were correct.

A final comment deserves attention, as many measurements are made with mixtures. In fact, the imposition of a temperature gradient in the fluid mixture generates a diffusive molecular/ionic flux, driven by one or more chemical potential/composition gradients, namely thermal diffusion or the Soret Effect. If in steady-state methods it can be assumed that it becomes zero and the Fourier Law defined the thermal conductivity, in transient methods, this is not obvious [19]. For a mixture of perfect gases, the same happens for the transient hot-wire method, and no mass transfer exists at the surface of the wire. Thus, we can assume that this result is kept for dense gases and liquids, and there no change in the working equation. Accordingly, Equations (10)–(13) can be applied to mixtures.

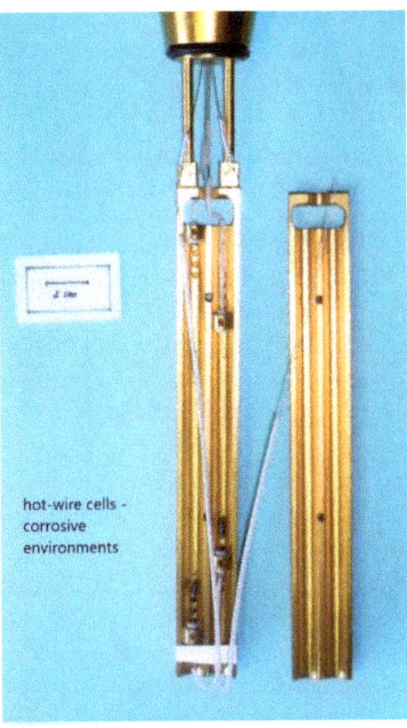

Figure 3. TiN film protection of TWH instrument for humid air measurements—detail of hot-wire cells. Hot-wire was made from tungsten (diameter 8.94 ± 0.05 μm, long wire length 148.92 ± 0.02 mm and short compensating wire length 71.69 ± 0.02 mm). The global uncertainty of the measurement, at a 95% confidence level, was estimated to be U_λ = 1.7%. Adapted from [35].

In principle, the hot-wire method can be applied to ionic melts and nanofluids, if a careful design is made of the instrument. However, the reliability and accuracy of instruments used to measure thermal conductivity of ionic melts has to consider several problems, which can influence the accuracy of the data obtained. Starting with the ionic melts, we have:

- High temperature measurements (molten salts).
- Radiation must be corrected or minimized, and convection eliminated.
- Electrical insulation of measuring probes indispensable.
- Use of current commercial instrumentation, such as transient hot-probes and transient hot-strips/disks, based on adaptations of the heat source immersed in the fluid exact behaviour, sacrifice accuracy at the cost of instrumentation and ease of handling of probes and samples. The accountability of methods must be proven.
- Many of the results presented in the bibliography were taken with commercial instrumentation, and not properly validated.
- In the limited context of hot wire, for instance, it is common to use just one wire, sometimes shorter than the model requires (end effects), which directly affects the accuracy of the measurements.

For the case of nanofluids, we can consider in addition that thermal conductivity is the property most influenced by the nanoparticles present. For a current analysis of the problem, see the recently published work of Antoniadis et al. [36]. In addition [11], the nanofluids can be electrically conducting (water, metal particles, nanocarbons, etc.) and when the ordinary transient hot-wire is applied to measure the thermal conductivity of electrical conducting liquids, several problems are encountered, as explained above, but can be summarized here:

- The contact between the bare metallic wire and the conducting liquid provides a secondary path for the flow of current in the cell and the heat generation in the wire cannot be defined unambiguously.

- Polarization of the liquid occurs at the surface of the wire, producing an electrical double-layer.
- The electrical measuring system (an automatic Wheatstone bridge) that detects the changes in the voltage signals in the wire is affected by the combined resistance /capacitance effect, caused by the dual path conduction.
- Insulated wires (necessity to cover the wire with materials or insulating anodizing coatings to protect it during the measurements) or polarized techniques are required [11].

Secondly, the generation of a high temperature gradient near the wire might create migration of nanoparticles to its surface and deposition of particles in the surface of the probes, creating nanostructures with organization dependent of surface of the probe/hot-wire, that vary the thermal conductivity of the non-perturbed nanofluid, giving rise to anomalous measured effects. It also changes its bulk concentration, distorting the propagation of the heating wave, making the sensing zone non-homogeneous, and originating a temperature discontinuity like the temperature jump in gases. A hot wire-nanofluid interface thermal conductance (sometimes referred to as the Kapitza resistance) needs to be measured, and this fact can be detected experimentally, as explained by Rutin and Skripov, 2017 [37] and Hasselman, 2018 [38]. Recent work in ionanofluids of graphene and carbon nanotubes in our group, using molecular simulation, showed that not only this interphase exists, but also that it has a finite value of thermal conductivity, of the order of 15–30% that of the bulk ionic liquid [39].

This is magnified when the nanoparticles have surface charges that create a problem equivalent to ions. This will affect the modelling of the measurement and will create erroneous results. These problems are not considered by instrument manufacturers, and especial probes must be designed on purpose. Tertsinidou et al. [40] have developed a comprehensive study of several nanofluids (ethylene glycol with added CuO, TiO_2, or Al_2O_3 nanoparticles and water with TiO_2 or Al_2O_3 nanoparticles or MWCNTs) and proved that the transient-hot wire method and the hot disk thermal analyser can produce comparable accurate results within 2% mutual uncertainty, if measurements are performed accordingly to the recommendations above. Their transient hot-wire cells are displayed in Figure 4. Figure 5 shows the results obtained compared with other authors.

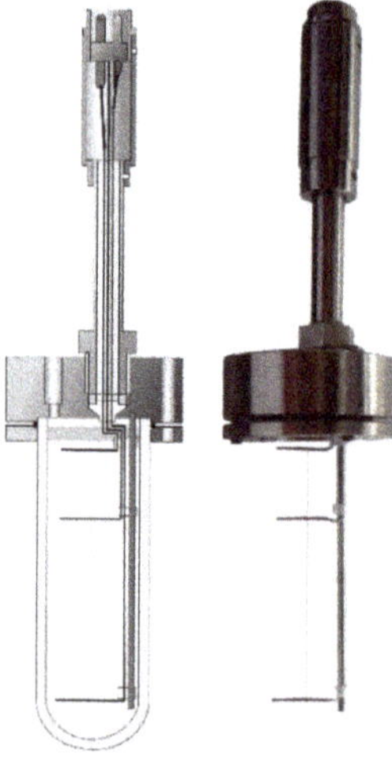

Figure 4. Hot-wire cells developed by Antoniadis et al. [36] for the measuring of the thermal conductivity of electrically conducting nanofluids. Tantalum (99.9%) wire (25 μm, long wire length 50 mm and short compensating wire length 20 mm) coated with an anodic layer of Ta_2O_5. The global uncertainty of the measurement, at a 95% confidence level, was estimated to be $U_\lambda = 1\%$. Courtesy of the authors.

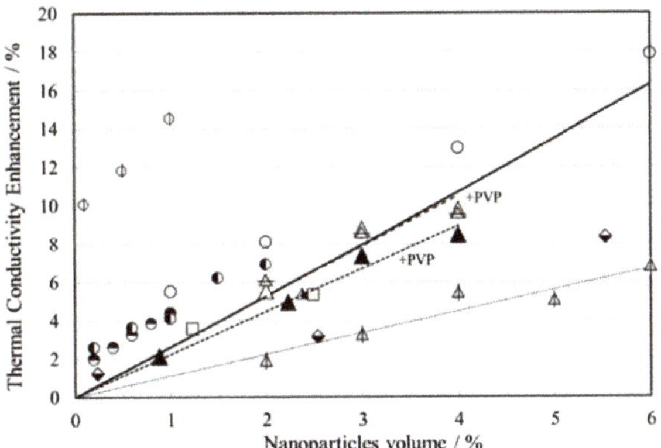

Figure 5. Thermal conductivity enhancement of water in the presence of TiO_2 nanoparticles as a function of the composition for nanoparticles with different diameters [40], at 298.15 K. Bottom-half-solid diamonds, 76 nm, Fedele et al. [41]; ϒ, 40 nm, Zhang et al. [42]; ◐, 26 nm, Duangthongsuk and Wongwises [43]; Φ, 25 nm, Yoo et al. [44]; ◓, 21 nm, Reddy and Rao [45]; ○, 21 nm, Yiamsaward et al. [46]; ▲, 20 nm, Haghighi et al. [47]; –, Hamilton–Crosser model [48]; ▲, 30–50 nm (+PVP), THW [40]; △, 5–15 nm (+PVP), THW [40]; triangle with vertical bar, 5 nm, THW [40]; triangle with horizontal bar, 5–15 nm (+PVP), HDTCA [40]. Reprinted with Permission from American Chemical Society (Further permissions related to the material excerpted should be directed to the ACS.).

In high temperature applications, like molten carbonates ($T > 600$ K) [49], the use of hot-wire geometries is limited, as very robust sensors are required. Geometries based on thin hot-wires cannot be used, due to their fragility. High temperature melts are solids at room temperature, have high surface tensions in the liquid phase, are strongly active, and both chemically and electrically conductive. Molten materials attack most materials either by chemical corrosion or physical erosion depending on the various environments, hence changing their chemical and physical properties. No known singular material satisfies all requirements and operation at high temperatures is always associated with some degree of interaction with the environment. So, any measuring probe must protect the sensing element from chemical or physical attacks, and therefore materials compatibility is a very challenging problem.

Until now, there have been no accurate experimental results with molten salts at very high temperatures. However, efforts to develop hot-strip sensors for high temperature in situ chemical gas streams have opened the way to their applications to molten salts and nanosalts (nanofluids of molten salts) [50].

Ceramic/thin metal film-based sensors, were designed, constructed, and applied for thermal conductivity measuring sensors, inspired in previous works of the group, by Lourenço et al. [51,52]. Platinum resistance thermometers were also developed using the same technology, to be used in the temperature measurement, were also constructed, and tested, using Al_2O_3 protecting electrically insulating coatings. The data acquisition system uses a linearization of the transient hot-strip model [53,54]. The model shows that the temperature rise at the surface of the hot strip is linearly proportional to the logarithm of a dimensionless time, τ, as displayed in Equations (14)–(16).

$$\tau = \frac{2\sqrt{a_f t}}{w} \quad a_f = \frac{\lambda}{\rho C_P} \tag{14}$$

$$\Delta T_s = T(t) - T_0 = \frac{q_s}{2\lambda\sqrt{\pi}} f(\tau) \tag{15}$$

$$f(\tau) = \tau \mathrm{erf}\left(\frac{1}{\tau}\right) - \frac{\tau^2}{\sqrt{4\pi}}\left[1 - \exp\left(-\frac{1}{\tau^2}\right)\right] + \frac{1}{\sqrt{4\pi}} E_1\left(\frac{1}{\tau^2}\right) \tag{16}$$

where a_f is the thermal diffusivity of the media, q_s is the heat dissipated per unit length of the strip, w is the width of the strip, τ, is a dimensionless time, and λ, ρ, $\tau_{,,}$ and C_P are, respectively, the thermal conductivity, density, and heat capacity at constant pressure of the media. The mathematical functions erf (y) and E_1 (z) are, respectively, the error function and the exponential integral, obtainable from rational approximations and easily programmed. Figure 6 shows one of the sensors.

Figure 6. Transient hot-strip sensor for high temperature thermal conductivity measurement. The sensor is composed by a metallic thin Platinum film, deposited by PVD on an alumina substrate. The hot strip lengths are l_L = 15 mm, l_S = 5 mm, and thickness w = 110 µm [49].

The dynamic response for a run with the sensor immersed in air can be seen in Figure 7, expressed as the temperature rise in the hot strip as a function of ln τ , with the optimal operating zone in red.

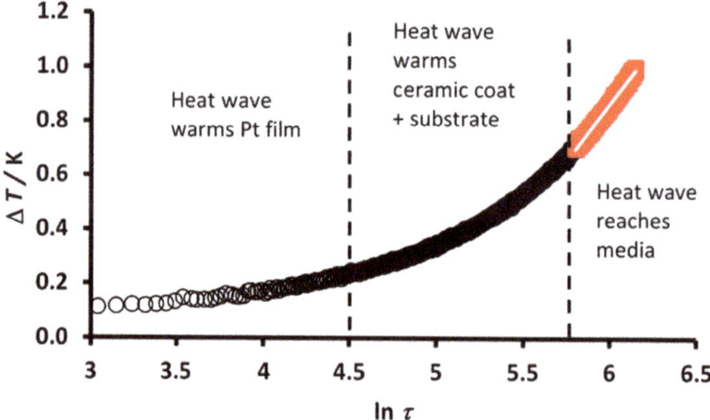

Figure 7. Dynamic sensor response for a run with the sensor immersed in air, with the optimal operating zone in red [49].

The equipment can measure the thermal conductivity of gaseous phases with an accuracy of 2–5% up to 840 °C, (95% confidence level). The extension of the sensors operating zone up to 1200 °C is under way, as are their assessment in high temperature melts.

3. Conclusions

The measurement of the thermal conductivity of ionic melts and nanofluids is proposed to be done with accurate methods of measurement, in order to have confidence in the quality of the results. The transient hot wire continues to be the best available technique to obtain good data, although methods like the transient hot strip, laser flash, hot disk, and photo-correlation spectroscopy (transparent fluids) might be a less accurate but cheaper method. Most equipment manufacturers sacrifice accuracy to easiness and speed of operation, and in most cases, produce bad data, which cannot be validated with standard reference data or certified reference materials. The art of measuring in thermal sciences is a continuously challenging field, especially when new and more complex systems for new applications appear. It is not enough to measure it, but rather to measure it well, and with an uncertainty compatible with the expected future use of the data.

Author Contributions: All the authors contributed equally to all manuscript. All authors have read and agreed to the published version of the manuscript.

Funding: This work was partially funded by FCT—Fundação para a Ciência e para a Tecnologia, Portugal, through Centro de Química Estrutural (UID/QUI/0100/2019).

Acknowledgments: This article is based upon work from COST Action Nanouptake, supported by COST (European Cooperation in Science and Technology) www.cost.eu. The authors would like to thank the members of this Action for several fruitful discussions along the years of 2017-19. The authors would also like to thank all the collaborators and partners in the field of thermal conductivity measurements, namely, J.M.N.A. Fareleira, U.V. Mardolcar, M.L.V. Ramires, A. Gurova, H.M. Roder, R.A. Perkins, M.J. Assael and, specially, to Sir W.A. Wakeham, who taught CANC the Art of Measuring Thermal Conductivity and the Tricks of the Trade.

Conflicts of Interest: The authors declare no conflict of interest.

References

1. Nieto de Castro, C.A. Absolute measurements of the viscosity and thermal conductivity of fluids. *JSME Int. J. Ser. II* **1988**, *31*, 387–401. [CrossRef]
2. Bird, R.B.; Stewart, W.E.; Lightfoot, E.N. *Transport Phenomena*, 2nd ed.; John and Wiley and Sons, Inc.: New York, NY, USA, 2002.
3. Wakeham, W.A.; Nagashima, A.; Sengers, J.V. *Measurement of the Transport Properties of Fluids: Experimental Thermodynamics*; Blackwell Scientific Pubs: Oxford, UK, 1991; Volume III.
4. Nieto de Castro, C.A.; Li, S.F.Y.; Nagashima, A.; Trengove, R.D.; Wakeham, W.A. Standard reference data for the thermal conductivity of liquids. *J. Phys. Chem. Ref. Data* **1986**, *15*, 1073. [CrossRef]
5. Mardolcar, U.V.; Nieto de Castro, C.A. The measurement of thermal conductivity at high temperatures. *High Temp. High Press.* **1992**, *24*, 551.
6. Assael, M.J.; Goodwin, A.R.H.; Vesovic, V.; Wakeham, W.A. *Experimental Thermodynamics Volume IX: Advances in Transport Properties of Fluids*; Especially Chapter 5; Royal Society of Chemistry: London, UK, 2014.
7. Nunes, V.M.B.; Queirós, C.S.; Lourenço, M.J.V.; Santos, F.J.V.; Nieto de Castro, C.A. Molten salts as engineering fluids—A review: Part I. Molten alkali nitrates. *Appl. Energy* **2016**, *183*, 603–611. [CrossRef]
8. Janz, G.J.; Allen, C.B.; Bansal, N.P.; Murphy, R.M.; Tomkins, R.P.T. *Physical Properties Data Compilations Relevant to Energy Storage*; NSRDS-NBS 61; Part II; USA Department of Commerce: Washington, DC, USA, 1979.
9. Nagasaka, Y.; Nagashima, A. The Thermal Conductivity of Molten $NaNO_3$ and KNO_3. *Int. J. Thermophys.* **1991**, *12*, 769–781. [CrossRef]
10. Nunes, V.M.B.; Lourenço, M.J.V.; Santos, F.J.V.; Matos Lopes, M.L.S.; Nieto de Castro, C.A. Accurate Measurement of Physico-Chemical Properties on Ionic Liquids and Molten Salts. In *Ionic Liquids and Molten Salts: Never the Twain*; Seddon, K.E., Gaune-Escard, M., Eds.; John Wiley & Sons, Inc.: New York, NY, USA, 2010; pp. 229–263.
11. Nieto de Castro, C.A.; Vieira, S.I.C.; Lourenço, M.J.V.; Sohel Murshed, S.M. Understanding Stability, Measurements, and Mechanisms of Thermal Conductivity of Nanofluids. *J. Nanofluids* **2017**, *6*, 804–811. [CrossRef]
12. Tada, Y.; Harada, M.; Tanigaki, M.; Egushi, W. Laser flash method for measuring thermal conductivity of liquids-Application to low thermal conductivity liquids. *Rev. Sci. Instrum.* **1978**, *49*, 1305–1314. [CrossRef]
13. Lee, S.W.; Park, S.D.; Kang, S.; Shin, S.H.; Kim, J.H.; Bang, I.C. Feasibility study on molten gallium with suspended nanoparticles for nuclear coolant. *Nucl. Eng. Des.* **2012**, *247*, 147–159. [CrossRef]
14. Kraft, K.; Lopes, M.M.; Leipertz, A. Thermal-diffusivity and thermal-conductivity of toluene by photon-correlation spectroscopy—A test of the accuracy of the method. *Int. J. Thermophys.* **1995**, *16*, 423–432. [CrossRef]

15. Nagasaka, Y.; Nakazawa, N.; Nagashima, A. Experimental determination of the thermal conductivity of molten alkali halides by the forced Rayleigh scattering method. I. Molten LiCl, NaCl, KCl, RbCl and CsCl. *Int. J. Thermophys.* **1992**, *13*, 555–574. [CrossRef]
16. Frez, C.; Diebold, G.J.; Tran, C.D.; Yu, S. Determination of the thermal diffusivities, thermal conductivities, and sound speeds of room-temperature ionic liquids by the transient grating technique. *J. Chem. Eng. Data* **2006**, *51*, 1250–1255. [CrossRef]
17. Bobbo, S.; Fidele, L. Experimental methods for the characterization of thermophysical properties of nanofluids. In *Heat Transfer Enhancement with Nanofluids*; Bianco, V., Manca, O., Nardini, S., Vafai, K., Eds.; CRC Press: Boca Raton, FL, USA, 2015; Chapter 4.
18. Nieto de Castro, C.A. State of Art in Liquid Thermal Conductivity Standards. In Proceedings of the Invited Communication to BIPM-CCT-WG9 Meeting, Bratislava, Slovak Republic, 9 September 2005.
19. Assael, M.J.; Nieto de Castro, C.A.; Roder, H.M.; Wakeham, W.A. Experimental Chemical Thermodynamics. In *Measurement of the Transport Properties of Fluids*; Wakeham, W.A., Nagashima, A., Sengers, J.V., Eds.; Blackwells: Oxford, UK, 1991; Volume 2, Chapter 7.
20. Ramires, M.L.V.; Nieto de Castro, C.A. Uncertainty and Performance of the Transient Hot Wire Method. In Proceedings of the Tempmeko 2001, 8th International Symposium on Temperature and Thermal Measurements in Industry and Science, Berlin, Germany, 19–21 June 2002; Fellmuth, B., Seidel, J., Scholz, G., Eds.; Vde Verlag GmbH: Berlin, Germany, 2002; Volume 2, pp. 1181–1886.
21. Beirão, S.G.S.; Ramires, M.L.V.; Dix, M.; Nieto de Castro, C.A. A New Instrument for the Measurement of Thermal Conductivity of Fluids. *Int. J. Thermophys.* **2006**, *27*, 1018–1041. [CrossRef]
22. Carslaw, H.S.; Jaeger, J.C. *Conduction of Heat in Solids*; Oxford University Press: London, UK, 1959; p. 256.
23. Healy, J.; de Groot, J.J.; Kestin, J. The Theory of the Transient Hot-wire Method for Measuring Thermal Conductivity. *Physica* **1976**, *82*, 392–408. [CrossRef]
24. Nieto de Castro, C.A.; Taxis, B.; Roder, H.M.; Wakeham, W.A. Thermal Diffusivity Measurement by the Transient Hot-Wire Technique: A Reappraisal. *Int. J. Thermophys.* **1988**, *9*, 293–316. [CrossRef]
25. Nagasaka, Y.; Nagashima, A. Absolute Measurement of the Thermal Conductivity of Electrically Conducting Liquids by The Transient Hot-Wire Method. *J. Phys. E Sci. Instrum.* **1981**, *14*, 1435–1440. [CrossRef]
26. Fisher, J. Zur Bestimmung der Wärmeleitfähigkeit und der Temperaturleitfähigkeit aus dem Ausgleichvorgang beim Schleiermacherschen Meßrohrverfahren und beim Plattenverfahren. *Ann. Phys.* **1939**, *34*, 669–688. [CrossRef]
27. Nieto de Castro, C.A. Medida da Condutibilidade Térmica de Hidrocarbonetos Líquidos pelo Método do Fio Aquecido. Ph.D. Thesis, Instituto Superior Técnico, UTL, Portugal, 1977.
28. Goldstein, R.J.; Briggs, D.G. Transient Free Convection about Vertical Plates and Circular Cylinders. *J. Heat Trans.* **1964**, *86*, 490–500. [CrossRef]
29. De Groot, S.R.; Kestin, J.; Sookiazian, H. Instrument to measure the thermal conductivity of gases. *Physica* **1974**, *75*, 454–482. [CrossRef]
30. Nagasaka, Y.; Nagashima, A. Simultaneous measurement of the thermal conductivity and thermal diffusivity of fluids. *High Temp. High Press.* **1989**, *21*, 363–371.
31. Menashe, J.; Wakeham, W.A.; Bunsenges, B. The Thermal Conductivity of n-Nonane and n-Undecane at Pressures up to 500 MPa in the Temperature Range 35–90 °C. *Phys. Chem.* **1981**, *85*, 340–347. [CrossRef]
32. Nieto de Castro, C.A.; Li, S.F.Y.; Maitland, G.C.; Wakeham, W.A. Thermal Conductivity of Toluene in the Temperature Range 35–90 °C at Pressures up to 600 MPa. *Int. J. Thermophys.* **1983**, *4*, 311–327. [CrossRef]
33. Calado, J.C.G.; Fareleira, J.M.N.A.; Nieto de Castro, C.A.; Wakeham, W.A. Reference state in thermal conductivity measurements. *Revista Virtual de Química* **1984**, *26*, 173–176.
34. Ramires, M.L.V.; Nieto de Castro, C.A.; Fareleira, J.M.N.A.; Wakeham, W.A. The Thermal Conductivity of Aqueous Sodium Chloride Solutions. *J. Chem. Eng. Data* **1994**, *39*, 186–190. [CrossRef]
35. Beirão, S.G.S.; Ribeiro, A.P.C.; Lourenço, M.J.V.; Santos, F.J.V.; Nieto de Castro, C.A. Thermal Conductivity of Humid Air. *Int. J. Thermophys.* **2012**, *33*, 1686–1703. [CrossRef]
36. Antoniadis, K.D.; Tertsinidou, G.J.; Assael, M.J.; Wakeham, W.A. Necessary Conditions for Accurate, Transient Hot-Wire Measurements of the Apparent Thermal Conductivity of Nanofluids are Seldom Satisfied. *Int. J. Thermophys.* **2016**, *37*, 78. [CrossRef]

37. Rutin, S.B.; Skripov, P.V. Comments on "The Apparent Thermal Conductivity of Liquids Containing Solid Particles of Nanometer Dimensions: A Critique" (Int. J. Thermophys. **2015**, *36*, 1367). *Int. J. Thermophys.* **2016**, *37*, 102. [CrossRef]
38. Haselman, D.P.H. Can the Temperature Dependence of the Heat Transfer Coefficient of the Wire—Nanofluid Interface Explain the "Anomalous" Thermal Conductivity of Nanofluids Measured by the Hot-Wire Method? *Int. J. Thermophys.* **2018**, *39*, 109. [CrossRef]
39. França, J.M.P.; Nieto de Castro, C.A.; Pádua, A.A.H. Molecular Interactions and Thermal Transport in Ionic Liquids with Carbon Nanomaterials. *Phys. Chem. Chem. Phys.* **2017**, *19*, 17075–17087. [CrossRef]
40. Tertsinidou, G.J.; Tsolakidou, C.M.; Pantzali, M.; Marc, J.; Assael Colla, L.; Fedele, L.; Bobbo, S.; Wakeham, W.A. New Measurements of the Apparent Thermal Conductivity of Nanofluids and Investigation of Their Heat Transfer Capabilities. *J. Chem. Eng. Data* **2017**, *62*, 491–507. [CrossRef]
41. Fedele, L.; Colla, L.; Bobbo, S. Viscosity and thermal conductivity measurements of water-based nanofluids containing titanium oxide nanoparticles. *Int. J. Refrig.* **2012**, *35*, 1359–1366. [CrossRef]
42. Zhang, X.; Gu, H.; Fujii, M. Experimental Study on the Effective Thermal Conductivity and Thermal Diffusivity of Nanofluids. *Int. J. Thermophys.* **2006**, *27*, 569–580. [CrossRef]
43. Duangthongsuk, W.; Wongwises, S. Measurement of temperature-dependent thermal conductivity and viscosity of TiO_2—Water nanofluids. *Exp. Therm. Fluid Sci.* **2009**, *33*, 706–714. [CrossRef]
44. Yoo, D.H.; Hong, K.S.; Yang, H.S. Study of thermal conductivity of nanofluids for the application of heat transfer fluids. *Thermochim. Acta* **2007**, *455*, 66–69. [CrossRef]
45. Reddy, M.C.S.; Rao, V.V. Experimental studies on thermal conductivity of blends of ethylene glycol-water-based TiO_2 nanofluids. *Int. Commun. Heat Mass Transf.* **2013**, *46*, 31–36. [CrossRef]
46. Yiamsawasd, T.; Dalkilic, A.S.; Wongwises, S. Measurement of the thermal conductivity of titania and alumina nanofluids. *Thermochim. Acta* **2012**, *545*, 48–56. [CrossRef]
47. Haghighi, E.B.; Saleemi, M.; Nikkam, N.; Khodabandeh, R.; Toprak, M.S.; Muhammed, M.; Palm, B. Accurate basis of comparison for convective heat transfer in nanofluids. *Int. Commun. Heat Mass Transf.* **2014**, *52*, 1–7. [CrossRef]
48. Hamilton, R.L.; Crosser, O.K. Thermal conductivity of heterogeneous two-component systems. *Ind. Eng. Chem. Fundam.* **1962**, *1*, 187–191. [CrossRef]
49. Nunes, V.M.B.; Lourenço, M.J.V.; Santos, F.J.V.; Nieto de Castro, C.A. Molten Alkali Carbonates as Alternative Engineering Fluids for High Temperature Applications. *Appl. Energy* **2019**, *242*, 1626–1633. [CrossRef]
50. Queirós, C.S.G.P.; Lourenço, M.J.V.; Vieira, S.I.; Serra, J.M.; Nieto de Castro, C.A. New Portable Instrument for the Measurement of Thermal Conductivity in Gas Process Conditions. *Rev. Sci. Instrum.* **2016**, *87*, 065105. [CrossRef]
51. Lourenço, M.J.V.; Serra, J.M.; Nunes, M.R.; Vallêra, A.M.; Nieto de Castro, C.A. Thin-Film Characterization for High Temperature Applications. *Int. J. Thermophys.* **1998**, *19*, 1253. [CrossRef]
52. Lourenço, M.J.V.; Nieto de Castro, C.A. Measuring the Thermal Conductivity at High Temperatures. Instrumental Difficulties and Sensor Performance. In Proceedings of the Tempmeko 2001, 8th International Symposium on Temperature and Thermal Measurements in Industry and Science, Berlin, Germany, 19–21 June 2002; Fellmuth, B., Seidel, J., Scholz, G., Eds.; Vde Verlag GmbH: Berlin, Germany, 2002; Volume 2, pp. 989–995.
53. Gustafsson, S.E.; Karawacki, E.; Khan, M.N. Transient hot-strip method for simultaneously measuring thermal conductivity and thermal diffusivity of solids and fluids. *J. Phys. D Appl. Phys.* **1979**, *12*, 1411–1421. [CrossRef]
54. Gustafsson, S.E.; Karawacki, E.; Khan, M.N. Determination of the thermal-conductivity tensor and the heat capacity of insulating solids with the transientmethod. *J. Appl. Phys.* **1981**, *52*, 2596–2600. [CrossRef]

Article

Entransy Dissipation Analysis and New Irreversibility Dimension Ratio of Nanofluid Flow Through Adaptive Heating Elements †

Fikret Alic

Faculty of Mechanical Engineering Tuzla, Department of Thermal and Fluid Technique, University of Tuzla, Tuzla 75000, Bosnia and Herzegovina; fikret.alic@untz.ba; Tel.: +387-(35)-320-939

† This article is an extended version of our paper published in 1st International Conference on Nanofluids (ICNf) and 2nd European Symposium on Nanofluids (ESNf), Castellón, Spain, 26–28 June 2019.

Received: 12 November 2019; Accepted: 19 December 2019; Published: 25 December 2019

Abstract: A hollow electric heating cylinder is inserted inside a thermo-insulating cylindrical body of larger diameter, together representing a single cylindrical heating element. Three cylindrical heating elements, with an independent electrical source, are arranged alternately one after the other to form a heating duct. The internal diameters of the hollow heating cylinders are different, and the cylinders are arranged from the largest to the smallest in the nanofluid's flow direction. Through these hollow heating cylinders passes nanofluid, which is thereby heated. The material of the hollow heating cylinders is a PTC (positive temperature coefficient) heating source, which allows maintaining approximately constant temperatures of the cylinders' surfaces. The analytical analysis used three temperatures of the hollow heating cylinders of 400 K, 500 K, and 600 K. The temperatures of the heating cylinders are varied for each of the three cylindrical heating elements. In the same arrangement, the inner diameters of the hollow cylinders are set to 15 mm, 11 mm, and 7 mm in the nanofluid's flow direction. The basis of the analytical model is the entransy flow dissipation rate. Furthermore, a new dimension irreversibility ratio is introduced as the ratio between entransy flow dissipation and thermal-generated entropy. This paper provides a suitable basis for optimizing the geometric and process parameters of cylindrical heating elements. An optimization criterion can be maximizing the new dimensionless irreversibility ratio, which implies minimizing thermal entropy and maximizing entransy flow dissipation.

Keywords: cylindrical heating element; nanofluid flow; entransy flow dissipation; thermal entropy

---

1. Introduction

In many electric heaters, whose function is based on the resistance generating heat, different fluids are used for heating. The shapes and dimensions of these heaters are different and depend on their application. If the heaters are made in the form of a hollow cylinder, then the fluid flows through its inner surface and, thus, is heated. The efficiency of heating a fluid along its length with standard in-line heaters varies from case to case. However, the overall efficiency and reliability of electric fluid heaters depend on several factors. The most common factors are fluid velocity, the shape and dimensions of the heating surfaces, and the temperature difference between the heating surfaces and the fluids. In this regard, researchers have been investigating different ways to maximize the efficiency of convective fluid heating. Fluids are often used for heating in heaters whose principle is based on thermal radiation. The total radiant power of the several heating elements and their infrared efficiency were studied in [1]. Furthermore, many studies investigated the effect of the internal structure of heating ducts on the efficiency of fluid heating. The pipe fitted with different rings in order to enhance the heat transfer rate between the pipe and the fluid was numerically investigated in [2]. Numerical investigations of the thermal and hydraulic characteristics of the turbulent convection of nanofluid flow in a pipe with conical ring inserts were conducted in [3]. A thermal system consisting of multiple electro-resistant storage elements has been investigated by several authors [4]. A mathematical model of the warm air heater and the performance of the thermo-electric warm air was investigated in [5]. Combining multiple electric resistance heating storage elements and maximizing their efficiency has been studied in [6]. The effect of the inactive zones in the cartridge heater on the temperature and heat flux distributions in the test cylinder was investigated in [7]. The combined effects of thermal radiation, friction dissipation, and the Joule heating flow of nanofluids were investigated in [8]. Additionally, many studies analyzed the different physical properties of nanofluids and their impact on enhancing thermal transfer performance [9–12]. The innovative process heater solution consists of several cylindrical heating elements. The total thermal entropy and outlet air temperature of the inline combination heating elements were introduced in [13]. This paper is based on the analytical analysis of the entransy flow dissipation rate. Some authors have investigated entransy dissipation [14–19]. The entropy generation in the flow of an electrically conducting couple stress nanofluid through a vertical porous channel subjected to constant heat flux was investigated in [20]. Entransy of an object is the heat transfer ability during a given time period, while entransy dissipation is essentially the thermomass energy dissipation during heat transfer. In this paper, three cylindrical heating elements are analyzed, with an independent electrical source, placed alternately one after the other, thus forming a sectional heating channel. Nanofluid flows through this sectional channel and, thus, heats up. The nanofluid analyzed consists of a base fluid (water) and Al_2O_3 nanoparticles.

Furthermore, the nanofluid is used as the working medium with different values of the nanoparticle volume fraction, flowing at different flow rates, through different, linearly connected heating elements. This paper is an extended conference paper [21], in which we introduced the new dimension irreversibility ratio. The entransy flow dissipation rate of three serially-connected cylindrical heating elements was analytically modeled and analyzed in this paper. The analytical analysis was conducted for the entransy flow dissipation rate, the new dimension irreversibility ratio, and the thermal resistance.

2. Materials and Methods

A hollow electric cylinder heater is inserted inside a thermo-insulating cylindrical body of larger diameter to form a single cylindrical heating element. Three cylindrical heating elements (Figure 1), with an independent electrical source, are positioned alternately one after the other to form a sectional heating duct. The inlet temperature of the nanofluid is 293 K, while its mass flow rate is varied. The temperatures of the three cylindrical elements had constant values of 400 K, 500 K, and 600 K, differently positioned in the nanofluid's flow direction. The cylindrical heating elements are relatively small, while the nanofluid flow is selected so that no evaporation of the nanofluid can occur as it passes through the three heating elements. Certainly, when using multiple heating elements, it is necessary to provide a ratio between the temperature of the heating elements and the flow rate to maintain a single-phase flow. In the analytical analysis, the temperatures of the three heating cylinders were 400 K, 500 K, and 600 K. In the same arrangement, the inner diameters of the hollow cylinders were 15 mm, 11 mm, and 7 mm in the nanofluid's flow direction. The cylindrical heat elements are made of PTC ceramic heating elements with insulating film. This PTC heater has many advantages, such as a constant temperature, a long lifetime, safe usage, and high efficiency for both low voltages (3–36 V) and high voltages (110–380 V).

The paper analyzes the entransy flow dissipation rate, thermal entropy generation, and thermal resistance for different positions of the cylindrical heating elements.

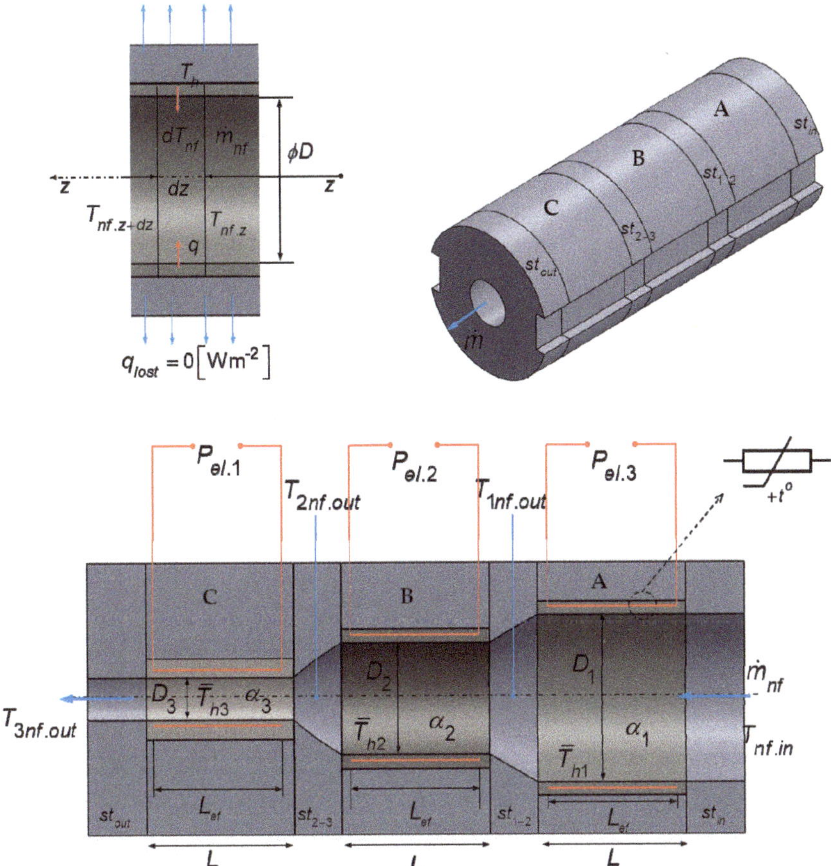

Figure 1. The in-line arrangement of the three hollow cylindrical heating elements.

In order to eliminate the effect of a sudden decrease in the cross-section on the pressure drop and the appearance of local turbulent nanofluid regions, a thermal insulation transition element was inserted between each heating element. In this way, the effect of the sudden change in the cross-section on the values of the local convective heat transfer coefficient can be neglected.

The effective length of the hollow heating element is less than the total length of the heating element, $L_{ef} < L$. This minimizes the influence of the nanofluid at the inlet and outlet of the heating element on the local efficiency of the surface of the cylindrical heating element. The thermal interaction between the hollow cylindrical heaters and the nanofluids can be represented by balance Equation (1):

$$\dot{m}_{nf} c_{nf} dT_{nf} = \alpha \left(T_h - T_{nf}\right) dA. \tag{1}$$

The density and specific heat capacity of the nanofluid are represented by Equations (2) [22], and (3) [23]:

$$\rho_{nf} = \frac{m_{np} + m_{bf}}{V_{np} + V_{bf}} = \frac{\rho_{np} V_{np} + \rho_{bf} V_{bf}}{V} = \rho_{np}\varphi + (1-\varphi)\rho_{bf} \tag{2}$$

$$c_{nf} = \frac{\rho_{np}\varphi c_{np} + (1-\varphi)\rho_{bf} c_{bf}}{\rho_{nf}} \tag{3}$$

where m_{np} and c_{np} are the mass and specific heat capacity of the nanoparticle, m_{bf} and c_{bf} are the mass and specific heat capacity of the base fluid. After grouping the physical sizes, we obtain:

$$\frac{dT_{nf}}{\left(T_h - T_{nf}\right)} = \frac{\alpha dA}{\dot{m}_{nf}c_{nf}} = \frac{\alpha D\pi\left[\rho_{np}\varphi + (1-\varphi)\rho_{bf}\right]}{\dot{m}_{nf}\left[\varphi\rho_{np}c_{np} + (1-\varphi)\rho_{bf}c_{bf}\right]}dz \tag{4}$$

where T_{nf} is the nanofluid's temperature, T_h is the temperature of the inner cylindrical element, φ is the nanoparticle volume fraction, ρ_{np} and c_{np} are the density and specific heat capacity of the particles,$\dot{m}_{nf}$ is the mass flow of the nanofluid, and ρ_{bf} and c_{bf} are the density and specific heat capacity of the base fluid. If the heat transfer between the surface of the heating elements and the environment is neglected, then from Equation (4) it follows that:

$$\int_{T_{nf.in}}^{T_{nf.out}} \frac{dT_{nf}}{\left(T_h - T_{nf}\right)} = \int_0^L \frac{\alpha D\pi\left[\rho_{np}\varphi + (1-\varphi)\rho_{bf}\right]}{\dot{m}_{nf}\left[\varphi\rho_{np}c_{np} + (1-\varphi)c_{bf}\rho_{bf}\right]}dz \tag{5}$$

where L is the length of one heating element and $T_{nf.out}$ and $T_{nf.in}$ are the outlet and inlet temperatures of the nanofluid. For the turbulent flow of nanofluid (Al_2O_3—water, 13 nm diameter spherical nanoparticles), the Nusselt number relation can be used [24]. In this case, the convective heat transfer coefficient is represented by Equation (6):

$$\alpha = \frac{Nu_{nf}\lambda_{nf}}{D} = \frac{0.021\mathrm{Re}_{nf}^{0.8}\mathrm{Pr}_{nf}^{0.5}}{D}\lambda_{nf} = \frac{0.021\left(\frac{w_{nf}D}{v_{nf}}\right)^{0.8}\left(\frac{v_{nf}\rho_{nf}c_{nf}}{\lambda_{nf}}\right)^{0.5}}{D}\lambda_{nf} = 0.021\frac{\left(w_{nf}\rho_{nf}\right)^{0.8}\left(c_{nf}\lambda_{nf}\right)^{0.5}}{D^{0.2}\mu_{nf}^{0.3}} \tag{6}$$

where the expression for dynamic viscosity [25] is:

$$\mu_{nf} = \frac{1}{(1-\varphi)^{2.5}}\mu_{bf} \tag{7}$$

while the nanofluid thermal conductivity [26] is:

$$\lambda_{nf} = \lambda_{bf}\frac{\left[\lambda_{np} + 2\lambda_{bf} + 2\left(\lambda_{np} - \lambda_{bf}\right)\varphi\right]}{\lambda_{np} + 2\lambda_{bf} - \left(\lambda_{np} - \lambda_{bf}\right)\varphi} \tag{8}$$

where λ_{np} and λ_{bf} are the thermal conductivity of the nanoparticle and the base fluid, respectively.

After integrating Equation (5), the outlet nanofluid temperature of the first cylindrical heating element is:

$$T_{nf1.out} = \overline{T}_{h1} - \left(\overline{T}_{h1} - T_{nf.in}\right)\cdot e^{-\frac{\alpha_1 \pi D_1 L\left[\rho_{np}\varphi + (1-\varphi)\rho_{bf}\right]}{\left[\varphi\rho_{np}c_{np} + (1-\varphi)\rho_{bf}c_{bf}\right]\dot{m}_{nf}}} \tag{9}$$

The resulting expression for the outlet temperature of the first heating element, Equation (9), can also be used for the other heating elements. Thus, in the case of three cylindrical heating elements (Figure 1), the fluid's outlet temperature of the first heating element represents the fluid's inlet temperature of the next heating element.

Equation (9) is used for each heating element, changing the internal diameters D_1, D_2, and D_3, and the convective heat transfer coefficients α_1, α_2, and α_3. Finally, the nanofluid's outlet temperature of the third element is shown by Equation (10):

$$T_{nf3.out} = \overline{T}_{h3} - \left\{ \begin{array}{l} \left(\overline{T}_{h3} - \overline{T}_{h2}\right) + \\ \left[\left(\overline{T}_{h2} - \overline{T}_{h1}\right) + \left(\overline{T}_{h1} - T_{nf.in}\right) \cdot e^{-\frac{\alpha_1 D_1 \pi L [\rho_{np}\varphi + (1-\varphi)\rho_{bf}]}{[\varphi \rho_{np} c_{np} + (1-\varphi)\rho_{bf} c_{bf}]\dot{m}_{nf}}} \right] \cdot \\ e^{-\frac{\alpha_2 D_2 \pi L [\rho_{np}\varphi + (1-\varphi)\rho_{bf}]}{[\varphi \rho_{np} c_{np} + (1-\varphi)\rho_{bf} c_{bf}]\dot{m}_{nf}}} \end{array} \right\} \cdot e^{-\frac{\alpha_3 D_3 \pi L [\rho_{np}\varphi + (1-\varphi)\rho_{bf}]}{[\varphi \rho_{np} c_{np} + (1-\varphi)\rho_{bf} c_{bf}]\dot{m}_{nf}}} \quad (10)$$

where T_{h1}, T_{h2}, and T_{h3} are the inner surface temperatures of the cylindrical heating elements. The lengths of all three heating elements are L, while their inner diameters decrease from D_1 to D_2 to D_3. Accordingly, the velocity of the nanofluid is different in each heating element and has the average values $w_{nf.1}$, $w_{nf.2}$, and $w_{nf.3}$, (Figure 1). The entransy flow dissipation rate [14], during nanofluid heating when the nanofluid passes through the three heating elements is:

$$\Delta\dot{E}_{ent.in-out} = \int_{nf.in}^{nf3.out} \dot{m}_{nf} c_{nf} T dT = \frac{1}{2}\dot{m}_{nf} c_{nf}\left(T^2_{nf3.out} - T^2_{nf.in}\right). \quad (11)$$

In order to mathematically connect the entransy flow dissipation rate and the thermal entropy, a new dimension irreversibility ratio (χ) is introduced, represented by Equation (12). This equation does not cover hydraulic-generated entropy resulting from hydraulic irreversibility in the flow of nanofluids through the heating elements.

$$\chi = \frac{\Delta\dot{E}_{ent.in-out}}{\dot{S}_{gen.\Delta T}} = \frac{\int_{nf.in}^{nf.out} \dot{m}_{nf} c_{nf} T dT}{\int_{nf.in}^{nf.out} \dot{m}_{nf} c_{nf} \frac{dT}{T}} = \frac{\frac{T^2}{2}\Big|_{nf.in}^{nf.out}}{\ln T\Big|_{nf.in}^{nf.out}} \quad (12)$$

Maximizing the new dimension irreversibility ratio χ is achieved by maximizing the entransy flow dissipation and minimizing the thermal-generated entropy. The new dimension irreversibility ratio χ is a function of two variables, entransy flow dissipation and thermal entropy, which can differentially affect the total change $\Delta\chi$. Derivation of Equation (12) yields:

$$d\chi = \frac{d\left(\Delta\dot{E}_{ent.in-out}\right)}{\dot{S}_{gen.\Delta T}} - \left[\frac{\Delta\dot{E}_{ent.in-out}}{\left(\dot{S}_{gen.\Delta T}\right)^2}\right] d\dot{S}_{gen.\Delta T} \quad (13)$$

or written as a relative ratio:

$$\frac{d\chi}{\chi} = \frac{d\left(\Delta\dot{E}_{ent.in-out}\right)}{\Delta\dot{E}_{ent.in-out}} - \frac{d\dot{S}_{gen.\Delta T}}{\dot{S}_{gen.\Delta T}}. \quad (14)$$

According to Equations (13) and (14), a relative change in $\Delta\chi/\chi$ can be determined if, during the heating process, the increases in the nanofluid's temperature are 20 °C, 40 °C, and 60 °C (Table 1).

Table 1. The relative change in the new dimension irreversibility ratio.

ΔT_{nf}[K]	$\frac{\Delta(\Delta\dot{E}_{ent.in-out})}{(\Delta\dot{E}_{ent.in-out})}$	$\frac{\Delta\dot{S}_{gen.\Delta T}}{\dot{S}_{gen.\Delta T}}$	$\frac{\Delta\chi}{\chi}$	$\frac{\Delta\chi}{\chi}100\%$
20	0.72	0.62	0.1	10
40	1.48	1.20	0.28	28
60	2.29	1.75	0.54	54

According to Table 1, the relative changes in entransy flow dissipation and thermal entropy are approximately the same. This provides a convenient opportunity to apply the optimization of Equation (14). Maximizing the relative change in the new dimension irreversibility ratio can be achieved by increasing the difference between the relative changes in entransy flow dissipation and thermal entropy. This maximization of $d\chi/\chi$ can be used to optimize different heat-fluidic devices. According to Equation (14), entransy flow dissipation and thermal entropy can be derived from ΔT_{nf} and written in a relative form; see Equations (15) and (16):

$$\frac{d\left(\Delta \dot{E}_{ent.in-out}\right)}{\Delta \dot{E}_{ent.in-out}} = \frac{\left(\Delta \dot{E}_{ent.in-out}\right)'}{\Delta \dot{E}_{ent.in-out}} d\left(\Delta T_{nf}\right) = \frac{2\left(1 + \frac{\Delta T_{nf}}{T_{nf.in}}\right)}{\left(2 + \frac{\Delta T_{nf}}{T_{nf.in}}\right)} \frac{1}{\Delta T_{nf}} d\left(\Delta T_{nf}\right) \tag{15}$$

$$\frac{d\dot{S}_{gen.\Delta T}}{\dot{S}_{gen.\Delta T}} = \frac{1}{T_{nf.in}\left(1 + \frac{\Delta T_{nf}}{T_{nf.in}}\right) \ln\left(1 + \frac{\Delta T_{nf}}{T_{nf.in}}\right)} d\left(\Delta T_{nf}\right) \tag{16}$$

In the case of three cylindrical heating elements, Equation (12) takes the form:

$$\chi = \frac{\Delta \dot{E}_{ent.in-out}}{\dot{S}_{gen.\Delta T}} = \frac{1}{2} \frac{\left(T^2_{nf3.out} - T^2_{nf.in}\right)}{\ln \frac{T_{nf3.out}}{T_{nf.in}}} = \frac{T^2_{nf.in}}{2} \frac{\left[\left(\frac{T_{nf3.out}}{T_{nf.in}}\right)^2 - 1\right]}{\ln \frac{T_{nf3.out}}{T_{nf.in}}} = \frac{T^2_{nf.in}}{2} \frac{\left(\gamma^2 - 1\right)}{\ln \gamma}. \tag{17}$$

On the other hand, the entransy dissipation number $\Delta\varepsilon$ presents the relationship between the actual entransy flow dissipation and the maximum entransy dissipation; see Equation (18). For a single heating element, this number $\Delta\varepsilon$ can be described as:

$$\begin{aligned} \Delta\varepsilon &= \frac{\Delta \dot{E}_{ent.in-out}}{\dot{Q}_{\max}\left(T_h - T_{nf.in}\right)} = \frac{1}{2} \frac{\dot{m}_{nf} c_{nf}\left(T^2_{nf.out} - T^2_{nf.in}\right)}{\dot{m}_{nf} c_{nf}\left(T_h - T_{nf.in}\right)^2} \\ &= \frac{T^2_{nf.in}\left[\left(\frac{T^2_{nf.out}}{T^2_{nf.in}}\right) - 1\right]}{2\left(T_h - T_{nf.in}\right)^2} \end{aligned} \tag{18}$$

where T_h is the heating element's surface temperature along its length. In the case of three heating elements with temperatures T_{h1}, T_{h2}, and T_{h3} and internal diameter D_1, D_2, and D_3, the entransy dissipation number is:

$$\Delta\varepsilon = \frac{\left(T^2_{nf3.out} - T^2_{nf.in}\right)}{2\left[\left(T_{h1} - T_{nf.in}\right)^2 + \left(T_{h2} - T_{nf1.out}\right)^2 + \left(T_{h3} - T_{nf2,out}\right)^2\right]}. \tag{19}$$

For three cylindrical heating elements, a comparison of the new dimensionless irreversibility ratio and the entransy dissipation number is described in the next equation:

$$\chi = \frac{\left[\left(T_{h1} \quad T_{nf.in}\right)^2 + \left(T_{h2} - T_{nf1.out}\right)^2 + \left(T_{h3} - T_{nf2,out}\right)^2\right]}{\ln \gamma} \Delta\varepsilon. \tag{20}$$

The total thermal resistance of the nanofluids when they pass through three regularly connected cylindrical heating elements is represented by Equation (21):

$$R_{ent} = \sum_{i=1}^{3} \frac{\Delta \dot{E}_{ent.i,in-out}}{\dot{Q}_i^2} = \sum_{i=1}^{3} \frac{1}{2} \frac{\dot{m}_{nf} c_{nf}\left(T_{nf.i}^2 - T_{nf.i-1}^2\right)}{\left[\dot{m}_{nf} c_{nf}\left(T_{nf.i} - T_{nf.i-1}\right)\right]^2} = \sum_{i=1}^{3} \frac{1}{2} \frac{\left(T_{nf.i} + T_{nf.i-1}\right)}{\dot{m}_{nf} c_{nf}\left(T_{nf.i} - T_{nf.i-1}\right)} \tag{21}$$

3. Results

By introducing some constraints, we analytically modeled the thermal interaction between the heating elements and the nanofluid's flow.

The nanofluid's mass flow, the temperature of the internal heating surface, and the nanoparticle volume fraction were varied. The temperatures of the internal heating surface had two arrangements: (1) 400 K, 500 K, and 600 K, and (2) 600 K, 500 K, and 400 K. The geometric arrangement of these heating elements is the same, with inner diameters of 15 mm, 11 mm, and 7 mm.

According to the analytical modeling carried out by using Equations (1)–(21), the following results were obtained. The characteristic physical sizes of the base fluid and nanoparticles are presented in Table 2. The same table shows the values of the geometric parameters according to Figure 1. For the nanofluid, Al_2O_3 was used for the basic nanoparticles, while the base fluid is water. The concentration of Al_2O_3 nanoparticles varied from 0.01 to 0.07, and to a maximum value of 0.3. The preparation method for Al_2O_3 nanoparticles is two-step dispersion in the base fluid.

For only one heating element, the inner diameter is 0.015 m, at different mass flows and temperatures of the internal heating surface. In this case, the ratio of the entransy flow dissipation rate for the nanofluid and the base fluid is shown in Figure 2.

Table 2. The physical properties of the nanoparticles and base fluid and the dimensions of the heating elements.

Substances	ρ (kgm^{-3})	λ (Wm^{-1}K^{-1})	c (Jkg^{-1}K^{-1})	D_1 (m)	D_2 (m)	D_3 (m)	L (m)
Al_2O_3	3970	40	791	0.015	0.011	0.007	0.015
Base fluid(water) at 60 °C	985	0.651	4184				

The entransy flow dissipation rate has a maximum value when the nanofluid is used, and this value increases linearly as the temperature of the inner surface of the cylindrical heating elements increases.

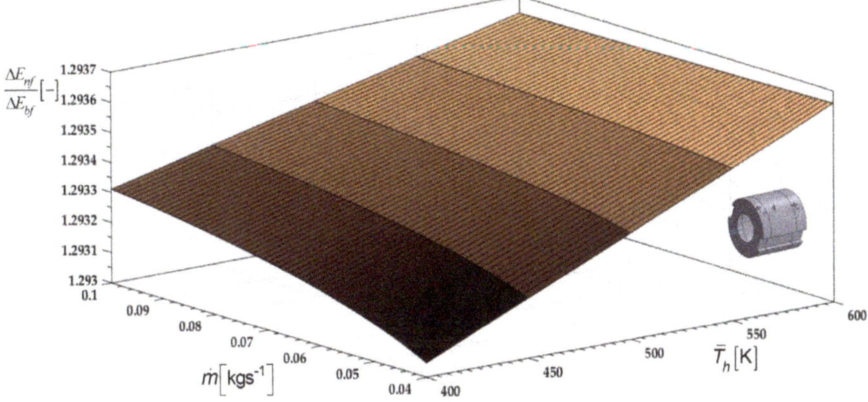

Figure 2. The dimensionless ratio of the entransy flow dissipation rate for one cylindrical heating element, with an inner diameter of 0.015 m.

Additionally, for the case of only one cylindrical heating element whose inside diameter is 0.015 m, the new dimension irreversibility ratio χ has high values of about 10^4 K^2 (Figure 3).

The value of this dimension ratio increases with increasing temperature of the inner surface of the cylindrical heating element. On the other hand, increasing the mass flow of this fluid decreases this ratio. As the fluid or nanofluid is retained for a shorter time in the cylindrical heating element, its outlet temperature and, therefore, the new dimension irreversibility ratio χ become lower.

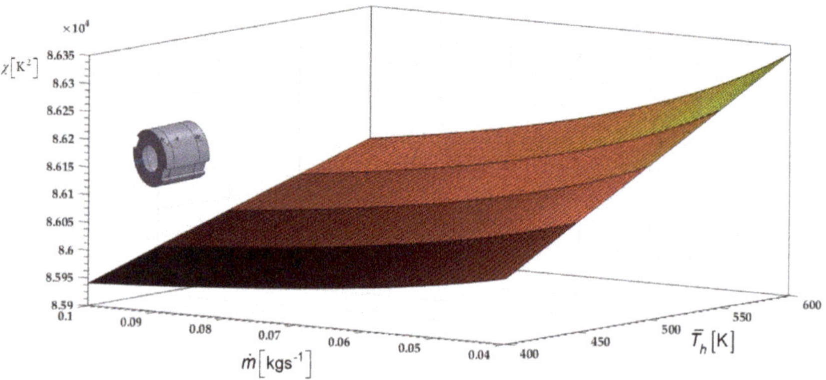

Figure 3. The new dimension irreversibility ratio for one cylindrical heating element.

In the case of using three cylindrical heating elements (Figure 1), the values of the entransy flow dissipation rate are shown in Figure 4. In the same figure, the temperatures of the heating elements and their internal diameters are varied. Thus, in the first case, the temperatures of the heating elements increase from 400 K to 500 K to 600 K, whereas, in the second case, the temperatures decrease from 600 K to 500 K to 400 K. As the nanoparticle volume fraction and the mass flow increase, the entransy dissipation rate increases (Figure 4). In the case of a temperature arrangement of 400 K, 500 K, and 600 K, the entransy flow dissipation rate increases faster.

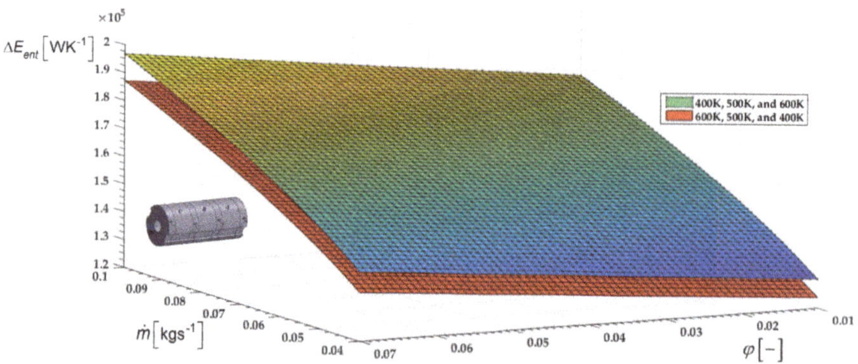

Figure 4. Effect of mass flow and the nanoparticle volume fraction on the entransy flow dissipation rate.

The use of nanofluids instead of the base fluid changes the values of the thermal resistances, as shown by their dimensionless relationship in Figure 5.

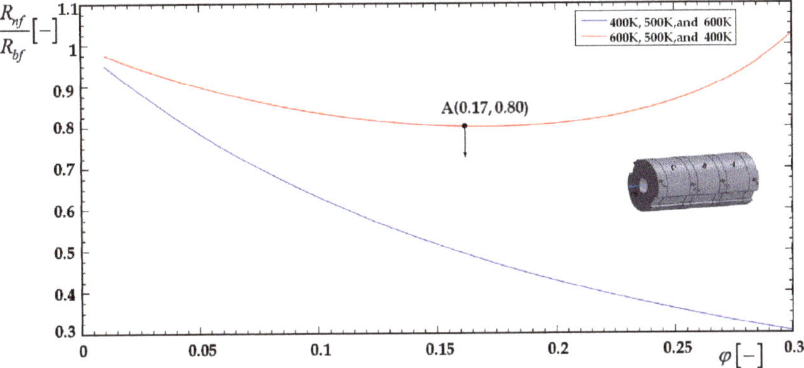

Figure 5. The thermal resistance ratio for a mass flow of 0.07 kg·s^{-1}.

The thermal resistance of the nanofluid is smaller than that of the base fluid. For heating element temperatures from 600 K to 500 K to 400 K, the minimum thermal resistance ratio is at a nanoparticle volume fraction of 0.17. The thermal resistances of the nanofluid and the base fluid, at heating element temperatures of 600 K, 500 K, and 400 K, are the same for the two nanoparticle volume fraction values of $\varphi = 0$ and $\varphi = 0.28$. In this case, the internal diameters of the heating elements and the temperatures of their surfaces are reduced: A (600 K, D = 15 mm), B (500 K, D = 11 mm), and C (400 K, D = 7 mm). At the same time, the nanofluid's temperature rises as it passes from heating element A to heating element C. The heat exchange between the heating elements and the nanofluids decreases, and thus the increase in nanoparticle volume fraction loses importance. After $\varphi = 0.17$, the R_{nf}/R_{bf} thermal resistance ratio approaches 1. Accordingly, the optimum value of the nanoparticle volume fraction for this case is 0.17. In the first case, for each heating element the following applies: A (400 K, D = 15 mm), B (500 K, D = 11 mm), and C (600 K, D = 7 mm), so there is a continuous increase in the heat exchanged between the heating elements and the nanofluid. The influence of the nanoparticle volume fraction rapidly reduces the thermal resistance of the nanofluid relative to the thermal resistance of the base fluid. With increasing nanoparticle volume fraction, the ratio χ_{nf}/χ_{bf} grows rapidly, especially for heating element temperatures of 400 K, 500 K, and 600 K (Figure 6).

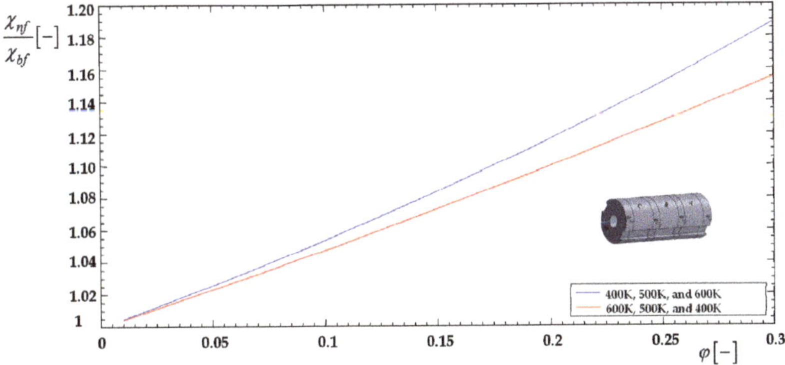

Figure 6. The new dimensionless irreversibility ratio for the nanofluid and base fluid, for a constant mass flow of 0.07 kg·s^{-1}.

The dimensionless ratio of the entransy dissipation number $\Delta\varepsilon_{nf}/\Delta\varepsilon_{bf}$ (Figure 7) shows the same qualitative trend as the new dimensionless irreversibility ratio $\Delta\chi_{nf}/\Delta\chi_{bf}$, as shown in Figure 6. On the right side, the ratio $\Delta\varepsilon_{nf}/\Delta\varepsilon_{bf}$ is greater than the ratio $\Delta\chi_{nf}/\Delta\chi_{bf}$. In both cases, the heating element arrangement of 400 K, 500 K, and 600 K shows a greater influence when nanofluid is used.

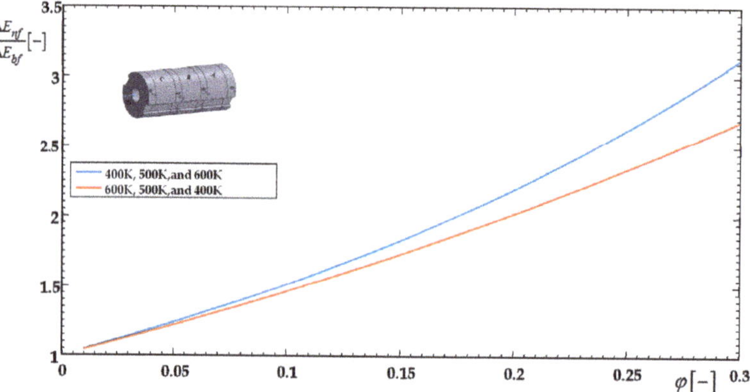

Figure 7. The ratio of the entransy dissipation number, for a mass flow of 0.07 kg·s^{-1}.

The dimensionless ratio of thermal resistance, R_{nf}/R_{bf}, as a function of nanoparticle volume fraction φ and mass flow is shown in Figure 8. This figure shows the individual influences of each of the three cylindrical heating elements, A, B, and C, which are mounted with inner diameters D_1 of 15 mm, 11 mm, and 7 mm in the flow direction. The lowest thermal resistance during the nanofluid flow is provided by heating element A (D_1 = 15 mm, 400 K). The lowest value of the ratio R_{nf}/R_{bf} is achieved for the highest nanoparticle volume fraction φ of 0.1.

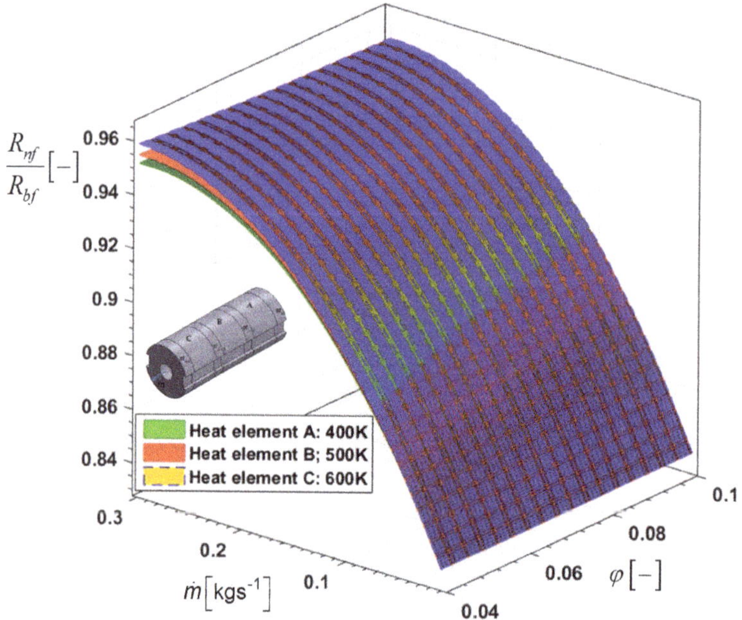

Figure 8. The ratio of the thermal resistance of individual heating elements, for arrangements of 400 K, 500 K, and 600 K.

If the cylindrical heating elements are kept in the same arrangement (15 mm, 11 mm, and 7 mm) in the flow direction and the inner surface temperatures are 600 K, 500 K, and 400 K, respectively, then the ratio R_{nf}/R_{bf} is as shown in Figure 9. As in the first case (Figure 8), the cylindrical heating element A shows the smallest influence of nanofluid on the thermal resistance ratio R_{nf}/R_{bf}.

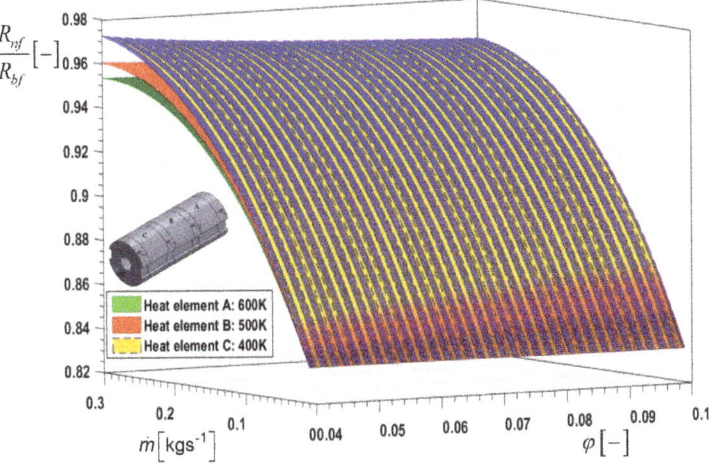

Figure 9. The ratio of thermal resistance of individual heating elements, for arrangements of 600 K, 500 K, and 400 K.

In order to verify the established mathematical model of nanofluid temperature change, when passing through three cylindrical heating elements, an experimental test was performed, Figure 10.

For the nanoparticles used Al_2O_3 while distilled water was used as the base fluid. An electric current regulator was also used to vary the constant temperatures of the heating PTC (positive temperature coefficient) elements. After the nanofluid temperature was measured, the same control setup was also used to measure the base fluid temperature, Figure 10.

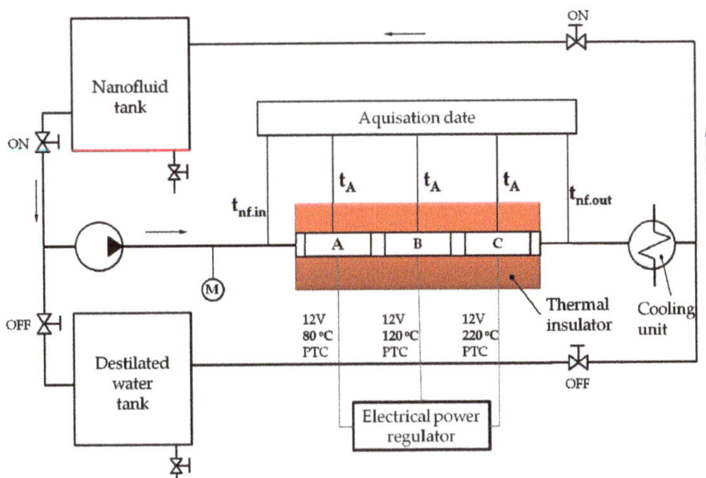

Figure 10. Experimental setup for testing the heating of nanofluids within heating elements.

Since the basic methodology in this paper is analytical modeling, experimental testing was used only to measure the rise in temperature of nanofluid and the base fluid. Indirect determinations of other physical quantities in this paper were not considered. Additionally, a comparative analysis of the obtained results was performed with the results obtained by analytical modeling. The following figures show a comparative analysis of the increase in water and nanofluid temperature with a nanoparticle volume fraction of 0.3, Figure 11, and 0.15, Figure 12.

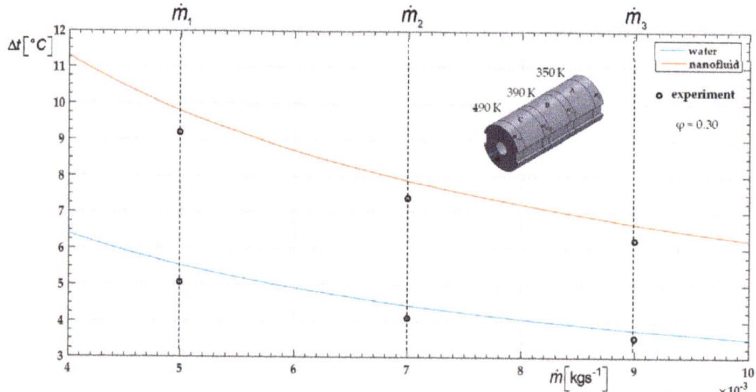

Figure 11. Increase of water and nanofluid temperature (φ = 0.3) through heating elements in the arrangement of 400 K, 500 K, and 600 K in flow direction.

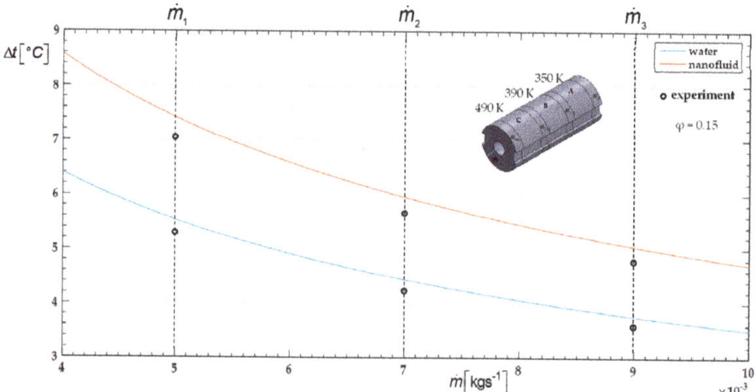

Figure 12. Increase of water and nanofluid temperature (φ = 0.15) through heating elements in the arrangement of 400 K, 500 K, and 600 K in flow direction.

4. Conclusions

For the selected heating system of three heating elements, with different temperatures of the internal surfaces and different internal diameters, we performed the analytical modeling of the characteristic thermal quantities of this system. Thermal interaction is achieved between each heating element and the nanofluid that flows through it. The selected nanofluid was Al_2O_3 for the basic nanoparticles, while the base fluid was water. The entransy flow dissipation rate of three serially connected cylindrical heating elements was analytically modeled and analyzed in this paper. Furthermore, a new dimension irreversibility ratio χ was introduced and analyzed using nanofluid and base fluid. The arrangement of the heating elements by their internal diameter was the same for all analyses, while only their temperatures were varied. The analytical analysis was conducted for the entransy flow dissipation rate, the new dimension irreversibility ratio, and the thermal resistance. The introduced new dimension irreversibility ratio connects entransy flow dissipation and thermal entropy. Nanofluid and base fluid were comparatively analyzed in this paper. The developed analytical model can be applied to various thermo-fluid processes. The new dimension irreversibility ratio can be used to optimize these processes. Maximizing this ratio implies simultaneously minimizing thermal entropy and maximizing the entransy flow dissipation rate, which can result in different optimal parameters.

Funding: The APC was funded by COST (European Cooperation in Science and Technology) Action CA15119: Overcoming Barriers to Nanofluids Market Uptake (NanoUptake).

Acknowledgments: This article is based upon work from COST Action Nanouptake, supported by COST (European Cooperation in Science and Technology) www.cost.eu.

Conflicts of Interest: The author declares no conflict of interest.

Nomenclature

D	Inner diameter of the heated cylindrical element (m)
L	Length of the heated cylindrical element (m)
$\dot{m}$	Mass flow (kg·s^{-1})
q	Heat flux (Wm^{-2})
A	Surface area (m^2)
P	Electric power (W)
φ	Nanoparticle volume fraction (-)
c	Specific heat capacity (Jkg^{-1}K^{-1})
ρ	Density (kg·m^{-3})
T	Temperature (K)
z	Coordinate (m)
α	Convective heat transfer coefficient (Wm^{-2}K^{-1})
V	Volume (m^3)
w	Velocity (ms^{-1})
Nu	Nusselt number (-)
Pr	Prandtl number (-)
Re	Reynols number (-)
λ	Thermal conductivity (Wm^{-1}K^{-1})
μ	Dynamic viscosity (Pa·s)
$\Delta\dot{E}_{ent}$	Entransy flow dissipation rate (WK^{-1})
χ	New irreversibility dimension ratio (K^2)
$\Delta\varepsilon$	Entransy dissipation number (-)
$\dot{S}_{gen.\Delta T}$	Thermal generated entropy (WK^{-1})
R_{ent}	Thermal resistance (KW^{-1})
Subscripts	
nf	Nanofluid
np	Nanoparticle
bf	Base fluid
h	Heater
st	Stabilizing element
1,2,3	First, second, third heated element
ent	Entransy
gen.ΔT	Thermal generated
nf.in	Inlet of nanofluid
nf.out	Outlet of nanofluid
max	Maximum
min	Minimum

References

1. Brown, K.J.; Farrelly, R.; O'Shaughnessy, S.M.; Robinson, A.J. Energy efficiency of electrical infrared heating elements. *Appl. Energy* **2016**, *162*, 581–588. [CrossRef]
2. Sripattanapipat, S.; Tamna, S.; Jayranaiwachira, N.; Promvonge, P. Numerical Heat Transfer Investigation in a Heat Exchanger Tube with Hexagonal Conical-ring Inserts. *Energy Procedia* **2016**, *100*, 522–525. [CrossRef]
3. Mohamed, H.A.; Abuobeida, I.A.M.A.; Vuthaluru, H.B.; Liu, S. Two-phase forced convection of nanofluids flow in circular tubes using convergent and divergent conical rings inserts. *Int. Commun. Heat Mass Transf.* **2019**, *101*, 10–20. [CrossRef]
4. Krane, R.J. *A Second Law Analysis of a Thermal Energy Storage System with Joulean Heating of the Storage Element*; WA/HT-19; ASME Paper: New York, NY, USA, 1985; Volume 85.
5. Menga, F.; Zhang, L.; Li, J.; Li, C.; Xie, L.; Luo, Y.; Liu, Z. Investigation of thermoelectric warm air heater. *Energy Procedia.* **2015**, *75*, 621–626. [CrossRef]
6. Bassily, A.M.; Colver, G.M. Modelling and performance analysis of an electric heater. *Int. J. Energy Res.* **2004**, *28*, 1269–1291. [CrossRef]
7. Chyu, M.C. Thermal analysis of the electrically heated cylindrical test section for heat transfer experiments. *Exp. Therm. Fluid Sci.* **1988**, *1*, 19–27. [CrossRef]
8. Daniel, Y.S.; Aziz, Z.A.; Ismail, Z.; Salah, F. Effects of thermal radiation, viscous and Joule heating on electrical MHD nanofluid with double stratification. *Chin. J. Phys.* **2017**, *55*, 630–651. [CrossRef]
9. Lebon, G.; Machrafi, H. A thermodynamic model of nanofluid viscosity based on a generalized Maxwell-type constitutive equation. *J. Non-Newtonian Fluid Mech.* **2018**, *253*, 1–6. [CrossRef]
10. Keblinski, P.; Phillpot, S.R.; Choi, S.U.S.; Eastman, J.A. Mechanisms of heat flow in suspensions of nano-sized particles (nanofluid). *Int. J. Heat Mass Transf.* **2002**, *45*, 855–863. [CrossRef]
11. Machrafi, H.; Lebon, G. The role of several heat transfer mechanisms on the enhancement of thermal conductivity in nanofluids. *Contin. Mech. Thermodyn.* **2016**, *28*, 1461–1475. [CrossRef]
12. Keblinski, P.; Eastman, J.A.; Cahill, D.G. Nanofluids for thermal transport. *Mater. Today* **2005**, *8*, 36–44. [CrossRef]
13. Alic, F. The non-dimensional analysis of nanofluid irreversibility within novel adaptive process electric heaters. *Appl. Therm. Eng.* **2019**, *152*, 13–23. [CrossRef]
14. Qian, X.; Li, Z. Analysis of entransy dissipation in heat exchangers. *Int. J. Therm. Sci.* **2011**, *50*, 608–614. [CrossRef]
15. Wang, H.; Liu, Z.; Wu, H. Entransy dissipation-based thermal resistance optimization of slab LHTES system with multiple PCMs arranged in a 2D array. *Energy* **2017**, *138*, 739–751. [CrossRef]
16. Wang, C.; Zhu, Y. Entransy analysis on optimization of a double-stage latent heat storage unit with the consideration of an unequal separation. *Energy* **2018**, *148*, 386–396. [CrossRef]
17. Alic, F. Entransy dissipation analysis of liquid vortex isolated by hollow cylinder. *Heat Transf. Res.* **2018**, *49*, 1689–1704. [CrossRef]
18. Liu, Y.K.; Tao, Y.B. Thermodynamic analysis and optimization of multistage latent heat storage unit under unsteady inlet temperature based on entransy theory. *Appl. Energy* **2018**, *227*, 488–496. [CrossRef]
19. Guo, Z.Y.; Zhu, H.Y.; Liang, X.G. Entransy—A physical quantity describing heat transfer ability. *Int. J. Heat Mass Transf.* **2007**, *50*, 2545–2556. [CrossRef]
20. Adesanya, S.O.; Ogunseye, H.A.; Falade, J.A.; Labelo, R. Thermodynamic Analysis for Buoyancy-Induced couple Stress Nanofluid Flow with Constant Heat Flux. *Entropy* **2017**, *19*, 580. [CrossRef]
21. Alic, F. Entransy dissipation analysis and new irreversibility dimension ratio of nanofluid flow through adaptive heating elements. In Proceedings of the 1st International Conference on Nanofluids (ICNf), Castello, Spain, 26–28 June 2019.
22. Wang, X.Q.; Mujumdar, A.S. A review on nanofluids—Part I: Theoretical and numerical investigations. *Braz. J. Chem. Eng.* **2008**, *25*, 613–630. [CrossRef]
23. Khanafer, K.; Vafai, K. A critical synthesis of thermophysical characteristics of nanofluids. *Int. J. Heat Mass Transf.* **2011**, *54*, 4410–4428. [CrossRef]
24. Park, B.C.; Cho, Y.I. Hydrodynamic and heat transfer study of dispersed fluids with submicron metallic oxide particles. *Exp. Heat Transf.* **1998**, *11*, 151–170.

25. Brinkman, H.C. The viscosity of concentrated suspensions and solutions. *J. Chem. Phys.* **1952**, *20*, 571. [CrossRef]
26. Maxwell, J.C.A. *Treatise on Electricity and Magnetism*, 2nd ed.; Clarendon Press: Oxford, UK, 1881.

MDPI
St. Alban-Anlage 66
4052 Basel
Switzerland
Tel. +41 61 683 77 34
Fax +41 61 302 89 18
www.mdpi.com

Energies Editorial Office
E-mail: energies@mdpi.com
www.mdpi.com/journal/energies

www.ingramcontent.com/pod-product-compliance
Lightning Source LLC
LaVergne TN
LVHW070658170726
843515LV00004B/437
9783039366620